SPRINGER
LABORATORY

W0232154

N. Latruffe   M. Bugaut   (Eds.)

Peroxisomes

With 14 Figures

Springer-Verlag Berlin Heidelberg GmbH

Professor Dr. NORBERT LATRUFFE
Dr. MAURICE BUGAUT

Universite de Bourgogne
Laboratoire de Biologie
Moleculaire et Cellulaire
21004 Dijon, BP 138
France

ISBN 978-3-642-87809-1   ISBN 978-3-642-87807-7 (eBook)
DOI 10.1007/ 978-3-642-87807-7

Library of Congress Cataloging-in-Puplication Data. Peroxisomes / N. Latruffe, M. Bugaut
(eds.). p. cm. - (Springer laboratory) Includes biliographical references and index. ISBN
3-540-56860-3 (alk. paper). - ISBN 0-387-56850-3 (alk. paper) 1. Peroxisomes-Congresses. I.
Latruffe, Norbert. II. Bugaut, M. (Maurice), 1945-. III. Series. [DNLM: 1.
Microbodies-congresses. QH 603.M35 P45352 1993] QH603.M35P46 1993   574.87'4--dc20
DNLM/DLC for Library of Congress 93-35650

© Springer-Verlag Berlin Heidelberg 1994

Originally published by Springer-Verlag Berlin Heidelberg New York in 1994 .

Typesetting: Camera ready by author
31/3145-5 4 3 2 1 - Printed on acid-free paper

Foreword

Since the characterization of peroxisomes by C. de Duve et al. in 1966, tremendous progress in peroxisome research has been made, especially over the past five years. This area has expanded so vastly and has become so diversified that it is now difficult to follow the weekly literature. Peroxisome research fields include metabolic function, biogenesis and protein translocation, molecular and clinical aspects of peroxisomal diseases, peroxisome proliferation by hypolipidemic drugs, plasticizers and herbicides, relationship between peroxisome proliferation, cell growth and hepatocarcinogenesis, gene organization and transcription, and the search for peroxisome proliferator nuclear receptors.

The growing interest in peroxisomes is illustrated by the fact that several learned societies have recently organized meetings on peroxisomes. However, no one has yet organized an advanced course including a practical laboratory session as well as basic lectures. The most recent comprehensive meetings on peroxisomes were organized in 1980 by Dr. P.B. Lazarow in New York and in 1986 by Dr. H.D. Fahimi in Heidelberg, so that the organization in 1993 of another international convention on peroxisomes would appear timely.

Peroxisome research is a field in which European laboratories appear to have a leading edge. This can be highlighted by the organization of an advanced course on peroxisomes held in Dijon, France from April 19-24, 1993, with the participation of European scientists who are at the forefront of peroxisomal study. The field has grown up rapidly in Dijon over the last few years, especially due to the fact that the GIS (Groupement d'Intérêt Scientifique) University-Industry in Cellular Toxicology has peroxisome proliferation as its focus. The organization of the FEBS advanced course 93-01 in Dijon, coupled to the SFBBM/SBCF (Société Française de Biochimie et Biologie Moléculaire/ Société de Biologie Cellulaire de France) joint international meeting, presents a subject of wide interest, not only for its basic biochemical and cell biological aspects, but also for its health implications (peroxisomal diseases; potential toxicity and carcinogenicity of peroxisome-proliferating drugs).

Sessions will be devoted to the presentation of the most recent findings on peroxisomes in the following areas: metabolic functions, morphology, biogenesis, proliferation and relationship to cell growth, and genetic diseases. The program, consisting of a practical course, lectures and discussions, should provide an excellent opportunity for cross-fertilization among the different basic and health science disciplines in the area of peroxisome research.

Dijon, May 1993   N. Latruffe and M. Bugaut

Preface

It is a pleasure to introduce this collection of the experiments and papers presented at the FEBS advanced course (practical session) on *Peroxisomes; Biochemistry, Molecular Biology and Genetic diseases* held in Dijon on April 19-24, 1993. The contents are an eloquent testimony to the advances accomplished in this field in recent years.

The peroxisome was born in Louvain in the early 1950s, together with its nonidentical twin, the lysosome, with which it happened to be closely associated in what was called at that time the light-mitochondrial fraction. The two siblings developed at very different paces. The lysosome almost immediately found its place in cell biology and entered human genetic pathology as early as 1963. In contrast, the peroxisome was a late bloomer. When Pierre Baudhuin and I first reviewed the topic in 1966, we still raised the possibility that the peroxisome might be a "fossil organelle." We immediately ruled out this possibility on evolutionary grounds, but the fact that it could be considered at all is evidence of how little we knew. Things then began to pick up. By the time Jim Hogg and I organized the first international meeting on peroxisomes at the New York Academy of Sciences in 1969, peroxisomes had emerged as ubiquitous organelles of eukaryotic cells, in animals, plants, fungi, and a variety of protists. Also, thanks to the work of Harry Beevers on the glyoxysomes of germinating castor bean seedlings, the role of peroxisomes in fatty acid metabolism was beginning to be appreciated. Clofibrate, identified morphologically as a peroxisome proliferator already in 1965, but recognized as such biochemically only ten years later, put the peroxisome on the pharmacological map. Finally, the discover by Sidney Goldfischer, in 1973, that peroxisomes are absent in the livers of patients with Zellweger's disease opened a new chapter of human genetic pathology, while at the same time providing cell biology with a unique model for the study of organellar biogenesis and assembly.

All these aspects of peroxisome biology and pathology, and many others, are represented by their latest developments in these proceedings, which will form a valuable addition to the growing literature on the Cinderella among cytoplasmic organelles.

Christian de Duve
Nobel Prize Winner

Contents

Part III: Toxicology and Pharmacology 119

Part IV: Cell Culture and Genetic Diseases 169

Supplement

Acknowledgements

The experiments described in this manual have been set up and managed by scientists of the following laboratories:

Laboratories   Experiments

Laboratoire de Biologie
Moléculaire et Cellulaire (LBMC)
Prof. Norbert LATRUFFE
Université de Bourgogne
BP 138 21004 Dijon, France   2, 4, 6, 7, 11, 12

Institüt für Anatomie und Zell Biologie II
Prof. H. Darius FAHIMI
Universität Heidelberg
Im Neuenheimer Feld 307
D-69120 Heidelberg, Germany   1, 3

Afdeling Farmacologie
Prof. Guy P. MANNAERTS
Katholieke Universiteit Leuven
Campus Gasthuisberg
B-3000 Leuven, Belgium   1

Laboratoire de Biochimie et
Toxicologie Alimentaires
Prof. Jean-Claude Lhuguenot
ENSBANA
1, place Erasme
21000 Dijon, France   8

Unité de Recherche: Génétique et
Mécanismes des Maladies du Foie de
l'Enfant
Dr. Jean DESCHATRETTE
INSERM U347
Hôpital de Bicêtre
80 Avenue du général Leclerc
94 276 Le Kremlin Bicêtre - Paris, France   13, 14

Laboratoire de Biochimie Pharmaceutique
Prof. Jean Charles TRUCHOT
Faculté de Pharmacie
Université de Bourgogne
BP 138 21004 Dijon, France   9

Unité de Toxicologie Nutritionnelle
Dr. Marc SUSCHETET
INRA
17, rue Sully
21000 Dijon, France   8

Central Toxicology Laboratory
Dr. Felix WAECHTER
CIBA GEIGY
R-1066 432 P.O. Box CH-4002
Basel, Switzerland   8

Central Toxicology Laboratory
Cellular and Molecular Biology
Dr. Stephen GREEN
I.C.I. ZENECA
Alderley Park, Macclesfield
Cheshire SK10 4TJ, England   5

Centre de Recherche de Daix
Dr. Bernard MAJOIE
Laboratoires FOURNIER
50, route de Dijon
21121 Fontaine-les-Dijon, France   10

Centre de Recherche en Toxicologie
Dr. Paul DESLEX
STERLING WINTHROP
5, bd Eiffel
BP 42 21602 Longvic-les-Dijon, France   2

Laboratoire d'Histologie, Section Toxicologie
Prof. Jean-Pierre ZAHND
Faculté de Médecine
Université de Bourgogne
BP 138 21004 Dijon, France   3

Special thanks to **M. Shawn J. HASSELL**, Research Assistant at the LBMC -
Dijon (a Sterling-Winthrop pre-doctoral fellowship) for English correction and
for the formatting of this Laboratory Manual., and **M. Courtney CHITWOOD**,
Technician at the LBMC - Dijon, for the typing of the manuscript.

The FEBS Practical Session was supported by the following companies and organizations: Alsace-Biologie, Amersham, Appligene, Beckman, Bioprobe, Biorad, Boehringer Mannheim France, Carlo Erba, Ciba Geigy, Consortium de Matériel de Laboratoire (C.M.L.), Conseil Régional de Bourgogne, Crédit Lyonnais, Dominique Dutscher, Dupont de Nemours - NEN, European Community, F.E.B.S. Advanced Course Organization, Flobio, Garnier, Gibco-BRL, Gilson Medical Electronics, IBM France, Isotopchim, René Janning, Jouan, Kontron, Laboratoires Fournier, Laboratoires Sterling-Winthrop, L'Air Liquide, Labsystems, Leica, Librairie de L'Université Flammarion, Merck, Mettler, Millipore, Oberval, Odil, Olympus Scop, OSI, Ozyme, Packard Instruments, Pharmacia, Polylabo, Prolabo, Promega, Roth, Sarstedt, Serva, Sigma-Aldrich, Schleicher et Schuell, Tebu, Techgen, Touzart et Matignon, Université de Bourgogne, Ville de Dijon - Services Municipaux.

List of Instructors

BARDOT Olivier, Doctoral Student
L.B.M.C.
Université de Bourgogne
BP 138 21004 Dijon, FRANCE

BAUMGART Eveline,
MD Instructor
Universität für Anatomie
und Zellbiologie(II)
Im Neuenheimer Feld 307
D-69120 Heidelberg, GERMANY

BENMBAREK Abedellah,
PhD Instructor
Groupe FOURNIER, Centre
de recherche de DAIX
60 route de Dijon 21121
Fontaine les Dijon, FRANCE

BENTEJAC Marc,
PhD Instructor
L.B.M.C.
Université de Bourgogne
BP 138 21004 Dijon, FRANCE

BOURNOT Paulette,
PhD Instructor
L.B.M.C.
Université de Bourgogne
BP 138 21004 Dijon, FRANCE

BREES Chantal, Technician
Katholieke Universiteit Leuven
Afdeling Farmacologie,
Campus Gasthuisberg
B-3000 Leuven, BELGIUM

BROCARD Cécile, Doctoral Student
L.B.M.C.
Université de Bourgogne
BP 138 21004 Dijon, FRANCE

BUGAUT Maurice,
PhD Instructor,
FEBS Practical Organizer
L.B.M.C.
Université de Bourgogne
BP 138 21004 Dijon, FRANCE

CAIRA Françoise, Doctoral Student
L.B.M.C.
Université de Bourgogne
BP 138 21004 Dijon, FRANCE

CAUSERET Catherine, Doctoral
Student, L.B.M.C.
Université de Bourgogne
BP 138 21004 Dijon, FRANCE

CHERKAOUI MALKI Mustapha,
PhD Instructor
L.B.M.C.
Université de Bourgogne
BP 138 21004 Dijon, FRANCE

CLEMENCET Marie-Claude,
PhD Instructor
L.B.M.C.
Université de Bourgogne
BP 138 21004 Dijon, FRANCE

CORNU-CHAGNON Marie-
Christine, PhD Instructor
Laboratoire de Biochimie
et Toxicologie alimentaires
ENSBANA, 1 Esplanade Erasme
21000 Dijon, FRANCE

GOUDONNET Hervé,
PhD Instructor
Laboratoire de Biochimie
Pharmacologique
Université de Bourgogne,
Bld Jeanne d'Arc
21000 Dijon, FRANCE

GREGOIRE Stéphane , Technician
Laboratoire de Biochimie
et Toxicologie alimentaires
ENSBANA, 1 Esplanade Erasme
21000 Dijon, FRANCE

LATRUFFE Norbert,
PhD Instructor, FEBS Coordinator
L.B.M.C.
Université de Bourgogne
BP 138 21004 Dijon, FRANCE

NG-BONAVENTURE Kim,
PhD Instructor
INSERM U347
Hôpital de Bicètre
80 Avenue du général Leclerc
94276 LE KREMLIN BICETRE
Paris, FRANCE

PACOT Corinne, Doctoral Student
L.B.M.C.
Université de Bourgogne
BP 138 21004 Dijon, FRANCE

SCHRADER Mickaël,
Doctoral Student
Universität Heidelberg; Institüt für
Anatomie und Zellbiologie (II)
Im Neuenheimer Feld 307
D-69120 Heidelberg, GERMANY

SIESS Marie-Hélène,
PhD Instructor
Unité de Toxicologie Nutritionnelle
Station Qualité des
Aliments de l'Homme
INRA, 17 rue de Sully
21000 Dijon, FRANCE

VAN VELDHOVEN Paul,
PhD Instructor
Katholieke Universität Leuven
Afdeling Farmacologie
Campus Gasthuisberg
B 3000 Leuven, BELGIUM

VÖLKL Alfred, PhD Instructor
Universität für Anatomie
und Zellbiologie (II)
Im Neuenheimer Feld 307
D-69120 Heidelberg, GERMANY

VONAU Marie Hélène, Technician
CIBA Geigy Ltd
R-1066 4 32 PO Box
CH-4002 Basle, SWITZERLAND

ZAHND Jean-Pierre, PhD Instructor
Laboratoire d'Histologie
Faculté de Médecine
Université de Bourgogne
BP 138 21004 Dijon FRANCE

I   Isolation and Characterization of Peroxisomes

General Indroduction to Isolation and Characterization of Peroxisomes

P.P. VAN VELDHOVEN and A. VÖLKL

Peroxisomes are ubiquitous cell organelles [1,4]. Most cells have spherical or spheroidal peroxisomes with a diameter between 0.2 and 1 μm, but in some tissues or organisms, like in the germinating spores of moss [8], in sebaceous glands [2], in mouse liver [3], in regenerating rat liver [10], and in the epithelium of kidney tubules [11], irregularly shaped profiles have been demonstrated. Peroxisomes are surrounded by a single membrane limiting the so called matrix compartment, which comprises the bulk of peroxisomal enzymes. Whereas most of these matrix enzymes are quite soluble, urate oxidase tends to aggregate, forming the crystalline "core" or "nucleoid" of peroxisomes.

In mammalia, peroxisomes are most abundant in liver and kidney. They are involved in catabolic as well as anabolic pathways, particularly those of cellular lipid metabolism, including the biosynthesis of ether lipids, cholesterol and dolichols and the degradation of fatty acids (and derivatives) via β-oxidation [6,7,9]. Physiologically important substrates for this last pathway include very long chain fatty acids, dicarboxylic fatty acids, prostaglandins, 2-methyl branched fatty acids, bile acid intermediates, and xenobiotics [6]. The first step in the peroxisomal breakdown of these substrates is catalyzed by an acyl-CoA oxidase. Peroxisomes contain also other oxidases which are involved in the degradation of purines, polyamines, D-amino acids, and 2-L-hydroxy acids. The generated hydrogen peroxide is destroyed by catalase, the most abundant enzyme of hepatic peroxisomes.

An important feature of hepatic and renal peroxisomes in rodents is their capability to respond to various xenobiotics with a massive proliferation. This process is usually accompanied by the induction of peroxisomal enzymes, particularly those of the β-oxidation of straight chain fatty acids [5].

In this experimental block, peroxisomes will be characterized biochemically and morphologically. The basic procedures to isolate peroxisomes will be demonstrated on rat liver. The purified peroxisomes will be further subfractionated, and the subfractions will be analyzed by immunoblotting to show the subperoxisomal localization of certain proteins. Measurements of typical peroxisomal enzymes (oxidases) will be performed in liver and kidney homogenates. Morphologically, peroxisomes will be visualized by electron microscopy in embedded tissues and fixed subcellular fractions, and by immunofluorescence in cultured cells.

References

1. Böck P, Kramar R, Pavelka M (1980) Peroxisomes and related particles in animal tissues. Cell Biol. Monographs. Vol 7. Springer-Verlag, New York
2. Gorgas K (1984) Peroxisomes in sebaceous glands. V. Complex peroxisomes in mouse preputial gland: serial sectioning and three-dimensional reconstruction studies. Anat Embryol 169:261 - 270
3. Gorgas K. (1985) Serial section analysis of mouse hepatic peroxisomes. Anat Embryol 172:21 - 32
4. Hruban Z, Rechcigl M Jr. (1969) Microbodies and related particles. Int Rev Cytol. Suppl. 1. Academic, New York
5. Lock EA, Mitchell AM, Elcombe CR (1989) Biochemical mechanisms of induction of hepatic peroxisome proliferation. Annu Rev Pharmacol Toxicol 29:145 - 163
6. Mannaerts GP, Van Veldhoven PP (1993) Metabolic role of mammalian peroxisomes. In: Peroxisome Importance in Biology and Medicine (Eds. Gibson and Lake), Taylor and Francis, London (in press)
7. Mannaerts GP, Van Veldhoven PP (1992) Role of peroxisomes in mammalian metabolism. Cell Biochem Funct 10:141 - 151
8. Pais MSS, Carrapico F (1982) Microbodies. A membrane compartment. Ann. N. Y. Acad Sci USA 306:510 - 513
9. Van den Bosch H, Schutgens RBH, Wanders RJA, Tager J (1992) Biochemistry of peroxisomes. Annu Rev Biochem 61:157 - 197
10. Yamamoto K, Fahimi HD (1987) Three-dimensional reconstruction of a peroxisomal reticulum in regenerating rat liver: evidence of interconnections between heterogeneous segments. J Cell Biol 105:713 - 722
11. Zaar K (1992) Structure and function of peroxisomes in the mammalian kidney. Eur J Cell Biol 59:233-254

1 Preparation and Purification of Peroxisomes; Subfractionation of Purified Peroxisomes; Acyl-CoA Oxidase Activity Measurements

P.P. VAN VELDHOVEN, C. BREES and A. VÖLKL

1.1 Isolation of Rat Liver Peroxisomes by Density Gradient Centrifugation

P.P. VAN VELDHOVEN and C. BREES

1.1.1 Introduction and Aims

The isolation of peroxisomes is generally accomplished in three steps. After homogenisation of the tissue or disruption of the cells, a subcellular fraction enriched in peroxisomes is obtained by differential centrifugation (for a comprehensive treatise on centrifugation, see [10]). This fraction is then further separated by means of density gradient centrifugation. Such a procedure will be demonstrated on liver, obtained from control rats and from rats treated with bezafibrate. Hypolipidemic drugs such as bezafibrate cause a pronounced proliferation of peroxisomes in liver and kidney of rodents. This is associated with hepatomegaly and the induction of a particular set of peroxisomal enzymes involved in the β-oxidation of fatty acids [8] (see Exp. 1.3).

Due to the apparent fragility of peroxisomes (cf. the presence of peroxisomal matrix proteins in the cytosolic fraction), mild conditions are employed during their isolation. The homogenate is normally prepared in an isotonic solution (0.25 M sucrose for mammalian tissues), which can be supplemented with various additives. Ethanol (0.1 % v/v) is often added to prevent the formation of inactive catalase compound II [7]. Low concentrations of buffers with a physiological pH, or sulphydryl protecting agents can be present as well. High salt concentrations should be avoided since these cause aggregation. The addition of chelators, especially during the further purification steps, tends to reduce the contamination of the purified peroxisomes with microsomal fragments.

By subjecting the homogenate to increasing g-forces, fractions are obtained that are enriched in a particular cell compartment. In a classical cell fractionation scheme, as developed by de Duve and coworkers for rat liver [2], five fractions are obtained: a nuclear, a heavy mitochondrial, a light mitochondrial, a microsomal and a cytosolic fraction. These are generally referred to as N, M, L, P and S-fractions.

The light mitochondrial fraction is enriched in lysosomes and peroxisomes. The size and shape of peroxisomes can vary much from species to species and from tissue to tissue. In rat liver and kidney, they are roughly spherical with an average diameter of 0.5 μm. In other tissues like brain and small intestine or in cultured cells, their diameter is considerably smaller (0.10 to 0.25 μm). Consequently, these peroxisomes (also called microperoxisomes) will sediment more slowly and will be recovered mainly in the microsomal fraction, but still contains a substantial amount of mitochondria and microsomal vesicles. Separation of peroxisomes from these contaminants is obtained by density gradient centrifugation. An ideal density gradient medium should fulfill following criteria [10]: be readily water-soluble; be inert to the biological material studied; possess known physico-chemical properties (high density, low viscosity, low diffusion constant, low osmotic activity); not interfere with the subsequent analytical procedures; be reasonably inexpensive. Sucrose meets most of these criteria, but the high osmolarity of dense solutions is a disadvantage. Moreover, the use of sucrose gradients requires pretreatment of the animals with Triton WR-1339 to lower the density of the lysosomes, which otherwise will contaminate the peroxisomal fraction [7]. With the newer iodinated gradient materials like Metrizamide (2-3-acetamido-5(N-methylacetamido)-2,4,6-tri-iodobenzoic acid; MW 789) [5,14] and Nycodenz (N,N'-bis (2,3-dihydroxypropyl)-5-N-(2,(3-dihydroxypropyl) acetamido-2,4,6-tri-iodo-isophthalamide; MW 821) [4,6,13], such pretreatment is not necessary. Separation and purification on the gradients made of these iodinated compounds is excellent, but the solutes are quite expensive.

Peroxisomes band in these gradients at a rather high density (1.22 to 1.25 g/ml) since their membranes are permeable to these low molecular weight substances [11,12]. If it is desirable to isolate peroxisomes under isotonic conditions, iso-osmotic Percoll gradients are most convenient. Percoll consists of 17 nm silica particles, coated with polyvinylpyrrollidone [9]. The purity of the final peroxisomal fraction is somewhat less, however [13].

By means of marker enzyme analysis (see [3,10]), one can calculate the purity of the final preparation, the yield and the enrichment over the homogenate. Reasonable enrichments are 30 to 40-fold. The total peroxisomal protein content of rat liver was previously estimated at 2.5 % [7] (peroxisomes isolated from Triton WR-1339 treated rats by means of sucrose gradients). By a refined isolation procedure this value was recently brought to 2 % [6], indicating that an enrichment of 50-fold is theoretically feasible.

For the practical demonstration, a Nycodenz step-gradient will be used. A printout with the basic procedures for Percoll and Metrizamide gradient centrifugation will be provided as well (see Addendum). The purified peroxisomes will be used in other sessions for the preparation of peroxisomal membranes (Exp. 1.2), immunoblotting of subperoxisomal fractions (Exp. 2), demonstration of the initial fixation and processing for electron microscopy (Exp. 3), and measurements of acyl-CoA oxidases (Exp. 1.3).

1.1.2 Equipment, Chemicals and Solutions

1.1.2.1 Equipment

Access to
- a guillotine or decapitation device
- 55 ml Potter-Elvehjem tissue grinders with loose fitting pestles (clearance 0.10-0.15 mm; Kontess ref. K-885510-0023)
- 15 ml Dounce tissue grinders with A and B pestles (Kontes ref. K-885300-0015)
- 7 ml Dounce tissue grinders with A and B pestles (Kontes ref. K-885300-0007)
- a refrigerated high speed centrifuge (Beckman J2-21) + a fixed angle 8 x 50 ml rotor (Beckman JA-20) + 50 ml capped polycarbonate tubes (Beckman PN 355669)
- a refrigerated ultracentrifuge (Beckman XL90) + a fixed angle 8 x 38.5 ml rotor (Beckman 70 Ti) + 26.3 ml thick wall polycarbonate bottles (Beckman PN 355618)
- a motor driven homogenization device (800 to 1500 rpm)
- a fraction collector + a calibrated one-channel peristaltic pump equipped with silicone tubing (i.d. ~ 1 mm; flow ~ 1 ml/min) and a long needle or canule (~ 12 to 20 cm; i.d. ~ 1 mm) + disposable 13-ml polystyrene tubes (~ 95 x 16 mm)
- balances

On the bench
- a timer
- containers with crushed ice
- tube racks (suitable to hold tubes of 30 mm diameter)
- glassware
- pipetting aid device (rubber bulb or motor driven)
- scissors
- a refractometer
- stirring rods to resuspend pellets in a gentle way; polystyrene rods made by Hellma (ref. 3001.21) or Kartell, originally designed to mix solutions inside photometer cuvettes, are convenient for this purpose.

1.1.2.2 Chemicals and Biological Material

- control and bezafibrate-treated rats (about 200 g)
- sucrose
- ethanol
- EDTA, Na_2 (ethylenediaminetetraacetic acid)
- MOPS ((3-[morpholino] propane) sulfonic acid)
- NaCl (p.a. grade)
- Nycodenz (anal. grade, Nycomed n° 227610)

1.1.2.3 Solutions

- 0.2 M MOPS (25 ml) brought to pH 7.2 with NaOH
- 0.1 M EDTA (10 ml) brought to pH 7.2 with NaOH
- homogenization solution (HM): 0.25 M sucrose - 5 mM MOPS pH 7.2 - 0.1 % (v/v) ethanol - 1 mM EDTA pH 7.2, freshly prepared (500 ml)
- Nycodenz solutions of 30 % (40 ml), 45 % (10 ml) and 56 % (w/v) (5 ml), containing 5 mM MOPS pH 7.2, 0.1 % (w/v) ethanol and 1 mM EDTA pH 7.2 (final concentrations); densities are 1.15, 1.24 and 1.30 g/ml, respectively.
- 0.9 % (w/v) NaCl (200 ml)

1.1.3 Experimental Procedure

1.1.3.1 Subcellular Fractionation

Due to the limited time for the demonstration, we will use a short fractionation procedure, in which we will first prepare a combined nuclear-heavy mitochondrial fraction, followed by sedimentation of lysosomes and peroxisomes (light mitochondrial fraction). Also due to time limits, we will not carry out further separation of the post-light mitochondrial supernatant into a microsomal and a cytosolic fraction.

Keep the following rules in mind: Cool media before use and keep them on ice; keep tubes, homogenizers, etc. on ice; precool rotors and centrifuges; store homogenate and subcellular fractions on ice; work as quickly and gently as possible; unless otherwise indicated, g-forces are calculated at the average radius (in mm) of the used rotor by means of the following equation : $g = (\text{rpm}/1000)^2 \times 1.12 \times r_{av}$; unless otherwise indicated, centrifugation time refers to the entire period from the start of the run (max. acceleration untill desired speed is obtained) untill the halt of the run (max. deceleration); always balance carefully the tubes in the rotor.

- Weigh the rat and decapitate; bleed and remove the liver; weigh the liver (compare the color and size of livers from control and fibrate-treated rats; calculate the amount of liver/100 g rat in order to obtain an idea of the hepatomegaly).
- Cut liver (not more than 10 g) in small pieces in a Potter tissue grinder containing 2 volumes of cold HM (keep about 1 g of liver aside and store it, together with the kidneys, in physiological saline on ice for the acyl-CoA oxidase measurements in the afternoon; Exp. 1.3).
- Homogenize tissue with 1 down-and-up stroke of a loose-fitting pestle rotating at about 1200 rpm.
- Pour the homogenate in a 50 ml centrifuge tube and spin at 1700 g for 12 min.

- Pour off the supernatant into a cylinder; resuspend the pellet in half of the original volume of HM; re-homogenize and spin again under the same conditions; repeat washing procedure once again; combine the supernatants of the washing procedures with the first supernatant and label the mixture E. The final pellet contains the unbroken cells, some red blood cells, large microsomal sheets, nuclei and the majority of the mitochondria.
- Bring the combined supernatant E to 60 ml per 10 g of starting liver; pour into centrifuge tubes and subject to 12300 g for 24 min.
- Decant the supernatant into a cylinder, gently resuspend the pellet in a small volume of HM by means of a Dounce homogenizer (15 ml; pestle A), bring to 1/2 of the original volume, and sediment again.
- Remove the supernatants by pipetting and combine (named PS); resuspend the pellet in a minimal volume of HM in a Dounce homogenizer (7 ml; pestle A) and adjust to 1 ml/2 g starting liver (named L).

1.1.3.2 Density Gradient Centrifugation

The purification of peroxisomes is normally performed on continuous density gradients. In the case of Nycodenz gradients, step gradients appear to result in comparable, if not better, separation and purification, and are easier to prepare [4,13].

For the demonstration, we will use a three-layer Nycodenz gradient consisting of a bottom-layer (56 % Nycodenz; d = 1.30 g/ml), which serves as a cushion, a middle layer (45 % Nycodenz; d = 1.24 g/ml) and an upper layer (30 % Nycodenz; d = 1.15 g/ml). The intact peroxisomes will band at the interface between the upper and the middle layer, while the majority of the contaminating organelles will not be able to enter the first layer. At the interface between the middle layer and the cushion, one can recover the free cores [13]. During homogenization, some peroxisomes apparently are damaged to such an extent that catalase can leak out and that the cores are released. These cores are rather heavy (d = 1.30 g/ml) and sediment mainly in the light-mitochondrial fraction, while the released catalase is recovered mainly in the cytosolic fraction. The (partially) empty peroxisomes or resealed peroxisomal membranes sediment together with microsomal constituents [1].

See the addendum for protocols using Percoll or Metrizamide gradients.

- Pipet in a 25 ml thick wall polycarbonate tube (Beckman PN 355618) 2 ml of 56 % Nycodenz, followed by 4 ml of 45 % and finally 16 ml of 30 % Nycodenz, and store tube in ice until use.
- Layer 2 ml (control rat) or 2 ml (fibrate rat) of the concentrated L-fraction onto the step-gradient.
- Centrifuge for 1 hr at 135,000 g, using a slow acceleration/deceleration mode.

- Remove the tubes from the rotor without disturbing the layers; insert a long needle, connected via tubing to a peristaltic pump, to the bottom of each tube. Start the pump (flow 1 ml/min) and collect fractions of 0.8 ml in pre-weighed 13 ml tubes.
- Determine the volume of collected fractions by measuring their weight and the refractive index of the solution. The density of the Nycodenz solution can be calculated from the refraction $[d = (3.242 \times \eta) - 3.323]$.

1.1.4 Results and Comments

The treatment of rats with bezafibrate (200 mg/kg body weight/day by stomach tube feeding) caused approximately a 1.5-fold increase in the liver size (3.80 ± 0.32 and 5.35 ± 0.15 g/100 g body weight in control and treated rats, respectively; means ± S.E.M.). This was accompanied by an 8 to 13-fold induction of the fibrate-inducible acyl-CoA oxidase (see sec. 1.3).

When enriched peroxisomal fractions, derived from the same amount of liver, were layered on the Nycodenz gradients, the larger number of peroxisomes (due to proliferation) in the treated rats was easily visible. The increase in peroxisomes with treatment is also reflected in the protein concentrations of the purified peroxisomes (see sec. 1.2).

A peculiar feature of peroxisomes from bezafibrate-treated rats is their tendency to adhere to the wall of the tube during gradient centrifugation, which might interfere with their purification, especially when using a fixed-angle rotor. Such behavior is apparently intrinsic to bezafibrate-induced peroxisomes, since it has also been observed during Metrizamide density gradient centrifugation. The phenomenon is almost absent, however, after clofibrate treatment.

References

1. Declercq PE, Haagsman HP, Van Veldhoven P, Debeer LJ, van Golde LMG, Mannaerts GP (1984) Rat liver dihydroxyacetone-phosphate acyltransferases and their contribution to glycerolipid synthesis. J Biol Chem 259:9064 - 9075
2. De Duve C, Pressman BC, Gianetto R, Wattiaux R, Appelmans F (1955) Tissue fractionation studies. 6. Intracellular distribution patterns of enzymes in rat liver tissue. Biochem J 60:604 - 617
3. Evans WH (1982) Subcellular membranes and isolated organelles: preparative techniques and criteria for purity. Techniques in Lipid and Membrane Biochemistry B407a: 1 - 46
4. Ghosh MK, Hajra AK (1986) A rapid method for the isolation of peroxisomes from rat liver. Anal Biochem 159:169 - 174
5. Hajra AK, Wu D (1985) Preparative isolation of peroxisomes from liver and kidney using Metrizamide density gradient centrifugation in a vertical rotor. Anal Biochem 148:233 - 244

6. Hartl FU, Just WW, Köster A, Schimassek H (1985) Improved isolation and purification of rat liver peroxisomes by combined rate zonal and equilibrium density centrifugation. Arch Biochem Biophys 237:124-134

7. Leighton F, Poole B, Beaufay H, Baudhuin P, Coffey JW, Fowler S, de Duve C (1968) The large-scale separation of peroxisomes, mitochondria and lysosomes from the liver of rats injected with Triton WR-1339. J Cell Biol 37:482 - 513

8. Lock EA, Mitchell AM, Elcombe CR (1989) Biochemical mechanisms of induction of hepatic peroxisome proliferation. Annu Rev Pharmacol Toxicol 29:145 - 163

9. Pertoft H, Laurent TC, Laas T, Kagedall L (1978) Density gradients prepared from colloidal silica particles coated with polyvinylpyrrolidone (Percoll). Anal Biochem 88:271 - 282

10. Rickwood D (1989) Centrifugation - A practical approach (2nd Ed.) Practical Approach Series. IRL Press, Oxford.

11. Van Veldhoven PP, Mannaerts GP (1993) Assembly of peroxisomal membranes. In: Eds. Maddy AH and Harris JR. Subcellular Biochemistry, membrane biogenesis. Plenum Publishing Press, New York (in press)

12. Van Veldhoven PP, Just WW, Mannaerts GP (1987) Permeability of the peroxisomal membrane to cofactors of β-oxidation: Evidence fro the presence of a pore-forming protein. J Biol Chem 262:4310 - 4318

13. Verheyden K, Fransen M, Van Veldhoven PP, Mannaerts GP (1992) Presence of small GTP-binding proteins in the peroxisomal membrane. Biochim Biophys Acta 1109:48 - 54

14. Völkl A, Fahimi HD (1985) Isolation and characterization of peroxisomes from the liver of normal untreated rats. Eur J Biochem 149:257 - 265

Addendum - Alternative Density Gradient Centrifugations

Isolation of Peroxisomes by Isotonic Percoll Gradients [12,13]

- Layer a peroxisome-enriched fraction (derived from 6 to 10 g of liver in 2 to 3 ml) on 36 ml (38.52 g) of a cold isotonic Percoll solution in a 40 ml centrifuge tube, and centrifuge for 1 hr at 35,000 g_{av} in a fixed-angle rotor, using slow acceleration/deceleration.

- Insert slowly a long needle, connected via tubing to a peristaltic pump, to the bottom and start to collect fractions (about 2.5 ml each). The first fraction contains the majority of lysosomes and free cores (responsible for the bimodal distribution of urate oxidase in Percoll gradients), and the two following fractions are enriched in mitochondria. Recover the microsomes (pinkish) at the top of the gradient and the peroxisomes (greenish) just below them.

- Prepare the Percoll solution as follows: to 20 g of Percoll (d $\pm$ 1.130; Pharmacia), add 15 ml of water, 5 ml of 10 mM EDTA.Na$_2$ (unbuffered), 3.77 g of sucrose, 50 μl of ethanol and, if necessary, 50 μl of 1 M dithiothreitol. Bring the solution to pH 8.0 by means of 20 mM MOPS solution (unbuffered, depending on the lot of Percoll, 1.5 to 3 ml needed), and to volume (50 ml) with water. This results in a final concentration of 20 % Percoll - 0.22 M sucrose - 0.1 % (v/v) ethanol - 1 mM DTT - 1 mM EDTA - $\pm$ 1 mM MOPS - pH 8.0.

- In the absence of EDTA, no separation is achieved between peroxisomes and microsomes. The addition of salts or working at a pH below 7.8 results in a bimodal distribution of catalase, apparently caused by co-sedimentation of some part of the peroxisomes with the mitochondria. By using unbuffered EDTA and MOPS solutions, the pH of the commercial Percoll solution is lowered without the addition of extra ions.

Isolation of Peroxisomes by Metrizamide Gradient Centrifugation

The approach is a modification of a procedure published a few years ago [14].
- Layer 5 ml of an enriched peroxisomal fraction (corresponding to ~ 5 g of liver) on top of a Metrizamide gradient contained in a 40 ml Quick-seal polyallomer tube and seal the tube using a commercial device (Beckman).
- Centrifuge in a vertical-type rotor (Beckman VTi50) at an integrated force of 1.265×10^6 g x min ($g_{max} = 39,000$), using slow acceleration/deceleration. Peroxisomes (greenish) band under the conditions employed in a density range of 1.23 - 1.24 g/ml.
- Recover fractions by means of a gradient collector or by aspiration using a syringe.
- To remove Metrizamide, which *e.g.* interferes with the determination of protein concentrations (Lowry method), dilute peroxisomal fractions about 10-fold with HM, followed by centrifugation (25,000 g x 20 min) to pellet the organelles.

Preparation of Metrizamide Gradients

The gradient used is formed as a step gradient transformed to a slightly exponential profile by subsequent freezing and thawing.

- Prepare Metrizamide solutions in 5 mM MOPS - 0.1 % (v/v) ethanol - 1 mM EDTA pH 7.2, adjusted to densities of 1.26, 1.225, 1.19, 1.155 and 1.12 g/ml, respectively.
- Layer 4,3,6,7 and 10 ml (1.26 - 1.12 g/ml) of these solutions sequentially to form a discontinuous gradient, which is subsequently frozen in liquid nitrogen and stored at -20°C.
- Prior to use, thaw the gradient as fast as possible to room temperature (about 30 min).

1.2 Subfractionation of Purified Rat Hepatic Peroxisomes and Isolation of Peroxisomal Membranes

A. VÖLKL

1.2.1 Introduction

Integral membrane proteins (IMP) of rat hepatic peroxisomes constitute about 10% of total peroxisomal proteins [1]. They are apparently synthesized on free polysomes [2,4,5], and are partly known to increase in rodent liver upon administration of peroxisome proliferators. Peroxisomal IMPs are involved in ether lipid biosynthesis, activation of very long-chain fatty acids and supposedly in ATP-driven transport processes into peroxisomes (see [6] for review). The 69/70 kDa polypeptide, one of the most prominent IMPs of peroxisomes, has been cloned and sequenced [3]. Most of the known membrane polypeptides, however, still have to be characterized biochemically as well as metabolically.

Isolated peroxisomes, albeit seemingly intact, are permeable to small molecules [7,8], yet the leakiness of the membrane most probably represents an isolation artefact. Consequently, however, in disassembling peroxisomes to prepare membranes, procedures such as osmotic shock to release matrix proteins are not very effective. In this session, the carbonate procedure of Fujiki et al. [1] will be employed to isolate peroxisomal IMPs. This approach is nondestructive, releases matrix and peripheral membrane proteins, and is the most straight-forward of the alternatives (*e.g.* freezing and thawing, pyrophosphate treatment, sonication) to subfractionate peroxisomes. However, the enzymatic activity of most of the proteins is lost upon this treatment.

1.2.2 Equipment, Chemicals and Solutions

1.2.2.1 Equipment

- an ultracentrifuge (Beckman XL90)
- a fixed angle rotor (Beckman 70.1 Ti, 12 x 13.5 ml) and polyallomer tubes (Beckman 326814)
- a swinging-bucket rotor (Beckman SW55 Ti, 6 x 5 ml) and polyallomer tubes (Beckman 326819)
- a spectrophotometer (Kontron, series UVIKON)
- a vortex mixer
- test tubes and racks
- automatic pipettes and tips
- Pasteur glass pipettes
- glassware
- disposable cuvettes (1 cm path length) for protein assay

1.2.2.2 Chemicals

- sucrose
- MOPS (Sigma)
- ethanol
- anhydrous EDTA (Sigma)
- sodium carbonate
- sodium hydroxide
- Bio Rad protein assay kit

1.2.2.3 Solutions

- homogenization medium (HM): 0.25 M sucrose, 5 mM MOPS, 0.1 % (v/v) ethanol, 1 mM EDTA, pH 7.2 (50 ml)
- 0.1 M Na_2CO_3, pH 11.5 (50 ml)
- 0.1 M NaOH (10 ml)

1.2.3 Experimental Procedure

1.2.3.1 Preparation of Integral Membrane Proteins

- Dilute as much as 1 ml of the peroxisome fraction freshly prepared from the livers of a control and bezafibrate-treated rat with HM to bring the total volume to ~12 ml (according to the volume of the tube fitting the fixed angle rotor): one tube per peroxisomal fraction.
- Centrifuge the 2 tubes in a fixed angle rotor at 40,000 g for 20 min.
- Decant the supernatants and rinse the pellets carefully with 1 ml HM each.
- Resuspend the pellets using a glass-rod in an appropriate volume of 0.1 M Na_2CO_3 to adjust protein concentration to 0.05 mg/ml .
- Keep the suspension on ice for 30 min.
- Centrifuge the tubes (2-4) in a swinging-bucket rotor for 1 h at 100,000 g.
- Decant the supernatants, which contain the matrix, core and peripheral membrane proteins, and save them.
- Rinse the pellets gently with ~1 ml of ice-cold water.
- Resuspend the pellets in 100 to 200 µl of an appropriate solution (HM, 0.1 M NaOH or sample buffer for SDS-PAGE).
- Divide in aliquots for protein determination, electrophoresis and immunoblotting.

Schematic Diagram:

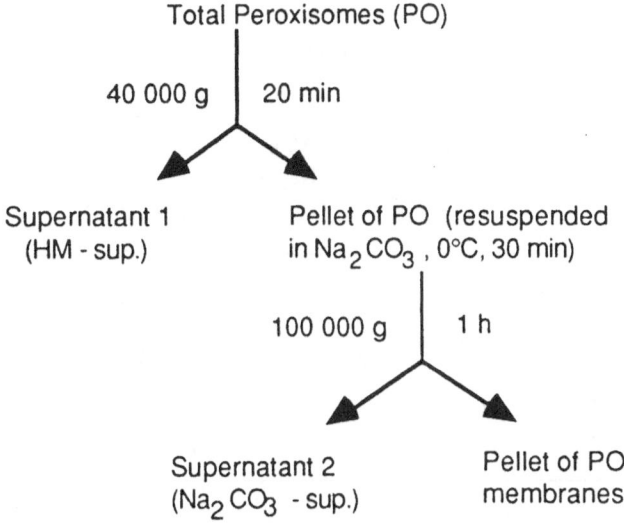

1.2.3.2 Protein Assay

Determine the protein concentrations in the total peroxisomal fractions as well as in the appropriate matrix and membrane subfractions according to the Bio-Rad microassay procedure.

1.2.4 Results

See Table 1.

1.2.5 Comments

- Steps 1-3 in the protocol "preparation of integral membrane proteins" are performed to concentrate the peroxisomal fractions for further processing and to remove a gradient medium such as Nycodenz.
- The pH of the carbonate solution is of crucial importance and should be kept at 11.0 - 11.5. Lower pHs are less effective or even ineffective [1].
- $K_2B_4O_7$ or K_2CO_3 may replace Na_2CO_3, whereas $NaHCO_3$, $NaCl$ or KCl cannot substitute for the carbonate even at a higher concentration than 100 mM [1].
- Peroxisomes obtained by Metrizamide gradient centrifugation should be freshly prepared for subfractionation by carbonate treatment. When stored frozen prior

to subfractionation, the isolated membranes proved to be contaminated with urate oxidase, which could not be removed by retreatment with carbonate.

Table 1. Results of a representative experiment

	Total PO	Sup. 1	PO in Na$_2$CO$_3$	Sup. 2	PO-Mb.
			mg protein (% recovery from total PO)		
Control rats	1.264 (100 %)	0.44 (34.7 %)	1.448	0.896 (70.9 %)	0.150 (11.8 %)
Bezafibrate-treated rats	4.972 (100 %)	1.23 (24.7 %)	3.056	2.464 (49.5 %)	0.550 (11.1 %)

Percentage recovery of proteins = μg protein in (Sup. 1 + Sup. 2 + PO-Mb.)(100)/(total PO)

Control:	118 %
Bezofibrate:	85 %

References

1. Fujiki Y, Fowler S, Shio H, Hubbard AL, Lazarow PB (1982) Polypeptide and phospholipid composition of the membrane of rat liver peroxisomes: Comparison with endoplasmic reticulum and mitochondrial membranes. J Cell Biol 93:103-110
2. Fujiki Y, Rachubinski RA, Lazarow PB (1984) Synthesis of a major integral membrane polypeptide of rat liver peroxisomes on free polysomes. Proc Natl Acad Sci 81:7127-7131
3. Kamijo K, Taketani S, Yokota S, Osumi T, Hashimoto T (1990) The 70-kDa peroxisomal membrane protein is a member of the Mdr (P-glycoprotein)-related ATP binding protein superfamily. J Biol Chem 265:4531-4540
4. Köster A, Heisig M, Heinrich PC, Just WW (1986) In vitro synthesis of peroxisomal membrane polypeptides. Biochem Biophys Res Comm 137:626-632
5. Suzuki Y, Ori T, Takiguchi M, Mori M, Hijikata M, Hashimoto T (1987) Biosynthesis of membrane polypeptides of rat liver peroxisomes. J Biochem 101:491-496
6. Van den Bosch H, Schutgens RBH, Wanders RJA, Tager JM (1992) Biochemistry of peroxisomes. Annu Rev Biochem 61:157-197
7. Van Veldhoven PP, Debeer LJ, Mannaerts GP (1983) Water- and solute accessible spaces of purified peroxisomes. Biochem J 210:685-693
8. Van Veldhoven PP, Just WW, Mannaerts GP (1987) Permeability of the peroxisomal membrane to cofactors of β-oxidation. J Biol Chem 262:4310-4318

1.3   Acyl-CoA Oxidase Activity Measurements

P.P. VAN VELDHOVEN and C. BREES

1.3.1 Introduction and Aims

Peroxisomes are biochemically defined as subcellular organelles containing hydrogen peroxide-decomposing catalase and hydrogen peroxide-producing oxidases. The latter act on different substrates (depending on the species : alcohols, (poly)amines, neutral and acidic D-amino acids, purine metabolites, L-2-hydroxy acids, acyl-CoAs) (see [3,6]). In the rat three acyl-CoA oxidases have been described so far [5,7] that are involved in the peroxisomal degradation of prostaglandins, dicarboxylic fatty acids, xenobiotics, very long chain fatty acids, bile acid intermediates, and branched chain fatty acids. The three enzymes have a characteristic substrate spectrum and tissue distribution [8]. Best known is palmitoyl-CoA oxidase, purified originally by the groups of Hashimoto [4] and Leighton [2]. It acts on the CoA-esters of long and very long chain fatty acids, short and long chain dicarboxylic acids and prostaglandins [8]. It has a native molecular mass of 139 kDa, is composed of 53 and 21.6 kDa subunits and is several-fold induced by treatment of rats with hypolipidemic drugs. Another oxidase, pristanoyl-CoA oxidase, acts on the CoA-esters of long chain 2-methylbranched fatty acids, like pristanic acid, but also on those of straight chain fatty acids [7,8]. Pristanoyl-CoA oxidase is not induced by peroxisome proliferators, is present in liver and kidney, possesses a molecular mass of 415 kDa (by gelfiltration) and consists of 70 kDa subunits [7]. A third acyl-CoA oxidase, trihydroxycoprostanoyl-CoA oxidase, is also not induced by proliferators and only found in liver. It acts on trihydroxycoprostanoyl-CoA, a bile acid intermediate, and is probably a dimer of 70 kDa subunits [5,8].

In this session, the different acyl-CoA oxidases will be measured in liver and kidney homogenates from control and fibrate-treated rats, in order to demonstrate their inducibility and their tissue distribution. The substrate-dependent hydrogen peroxide production will be quantified by means of a peroxidase-catalyzed dimerisation of homovanillic acid into a fluorescent product [1]. Although different hydrogen donors have been proposed, we prefer the use of homovanilic acid since it is very stable and results in a high sensitivity. In addition, these fluorimetric assays can be applied to all oxidases and, due to the small volumes, only low amounts of (sometimes) costly subtrates are consumed. However, many pitfalls can be encountered, as summarized in Fig. 1. The produced hydrogen peroxide can be destroyed chemically by sulfhydryl compounds (glutathione, dithiothreitol) or enzymatically by catalase. The interference of the latter one can be reduced by preincubating the samples with azide. If the enzyme is stable, chemical interference can be removed by dialysis, or, if the activity is high, by dilution. If the enzyme is not affected, pretreatment of the samples with

N-ethylmaleimide removes the sulphydryl interference. The presence of H-donors in sufficient amount to compete with homovanillic acid (NADH, ascorbic acid, serotonin, uric acid) in biological samples is less likely. With CoA-esters, one encounters another problem: hydrolysis (either chemically in the stock solutions or enzymatically during the assay) generating free CoA-SH, which combines with hydrogen peroxide. The chemical interference is responsible for the lag phase in the production of peroxide when analysing crude homogenates, crude nuclear fractions and cytosolic fractions.

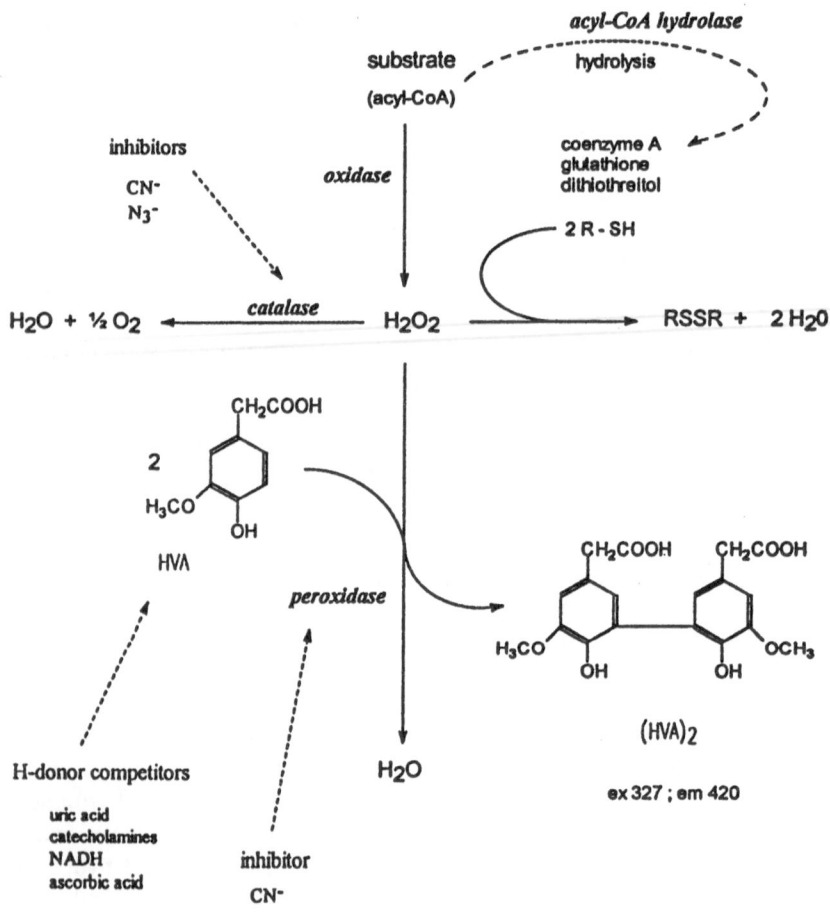

Figure 1: Possible pitfalls in fluorometric analysis. *HVA* : homovanillic acid; *(HVA)2* : homovanillic acid dimer.

1.3.2. Equipment, Chemicals and Solutions

1.3.2.1 Equipment

Access to
- a motor driven homogenization device (800 to 1500 rpm)
- a spectrofluorimeter (preferentially with continuous spectrum adjustment; no filters; Kontron SFM 25) + quartz cuvettes of 4 ml of less (10 mm optical path)
- an eppendorf centrifuge (rotor for 24 1.5 ml tubes)
- an automatic dispensing device (delivering between 1 and 3 ml)
- a thermostated shaking water bath at 37 °C
- a vortex
- 55 ml Potter-Elvehjem tissue grinders with loose fitting pestles (clearance 0.10-0.15 mm; Kontes ref. K-885510-00223)(see experiment 1.2) and a tight fitting pestle (has to be custom-made).

On the Bench
- pieces of cotton wool gauze (~ 15 x 15 cm) + glass or plastic funnels
- timers
- Eppendorf tubes of 1.5 ml + racks
- glass or plastic reaction tubes of 14 ml (~ 100 x 16 mm) + racks
- glass or plastic reaction tubes of 5 ml (~ 75 x 12 mm) + racks
- adjustable automatic pipettes + tips
- colored felt pens (blue, red, green)

1.3.2.2 Chemicals and Biological Material

- pieces of rat liver from control and bezafibrate-treated animals and kidneys from control rats (see experiment 1.2)
- homovanillic acid (HVA; 98%; Aldrich ref. 14, 364-2)
- peroxidase (POD; horse radish, grade II, ~200 U/mg, Boehringer ref. 127361; 10,000 U)
- palmitoyl-CoA (Pharmacia)
- 2-methylpalmitoyl-CoA (not commercially available)
- trihydroxycoprostanoyl-CoA (THC-CoA; not commercially available)
- uric acid (Merck ref. 817)
- uricase (hog liver, 2 mg/ml; Boehringer ref 127469)
- flavine adenine nucleotide (FAD; Boehringer ref. 1102338; 200 mg)
- sodium carbonate and sodium bicarbonate
- EDTA, Na_2
- perchloric acid
- sodium azide (NaN_3)
- defatted bovine serum albumin (BSA)

1.3.2.3 Solutions

- 50 μM uric acid (weigh ~1 mg uric acid; dissolve in 100 ml of 50 mM K-phosphate buffer pH 7.5; n.easure absorbance at 293 nm (ε = 12,600) and bring to 50 μM; store at -20°C in aliquots)
- 1 mg/ml POD (dissolve 10 mg in 10 ml of water; store at -20°C in aliquots)
- 60 mM HVA (dissolve 55 mg in 10 ml of 0.1 N HCl; store at -20°C in aliquots; warm before use to remove some haziness)
- 100 mM NaN$_3$ (10 ml; store frozen)
- 1 mM FAD (1 ml; store frozen and shielded from light)
- 6 % (w/v) BSA (5 ml)
- 0.2 M K-phosphate buffer - pH 8.3 (100 ml)
- 1 mM palmitoyl-CoA (store frozen in aliquots; A_{260} = 15.4; 3.5 ml)
- 1 mM 2-methylpalmitoyl-CoA (store frozen in aliquots; A_{260} = 15.4; 3.5 ml)
- 1 mM THC-CoA (store frozen in aliquots; A_{260} = 15.4; 3.5ml)
- homogenization medium (HM; see Exp 1.2) (100 ml)
- HClO$_4$ 8 % (w/v)
- 0.5 M carbonate buffer pH 10.7 (bring 600 ml of a 0.5 M Na$_2$CO$_3$ solution containing 10 mM EDTA to pH with approx 60 ml of a 0.5 M NaHCO$_3$ solution containing 10 mM EDTA)
- urate oxidase (dilute Boehringer solution 3 fold in water just before use)

1.3.3 Experimental Procedure

- Label 3 sets of 6 reaction tubes of 5 ml (A to F) in green, blue and red.
- Label 3 sets of 24 Eppendorf tubes (A to F with index corresponding to stop-times indicated in Table 2) in green, blue and red.
- Prepare 10 % (w/v) homogenates of liver and kidney in HM (5 x up-and-down strokes of loose and tight fitting pestles) and filter through cotton wool gauze.
- Prepare reaction mixtures (see Table 2).
- Dilute homogenates with HM according to Table 2.
- Place 80 μl of diluted homogenates in the bottoms of 5 ml reaction tubes (marked A to F), followed by 20 μl of FAD-NaN$_3$ (see Table 2), incubate at 0°C for 5 to 10 min, and start reaction by adding 400 μl of reaction mix (see Table 2 for composition); remove at indicated times 100 μl, which is added to the labeled Eppendorf tubes containing 40 μl of HClO$_4$.
- Centrifuge the Eppendorf tubes, mix 100 μl of supernatant with 1.4 ml of carbonate buffer and read fluorescence after ~10 min (ex 327 nm - slit 2 nm; em 420 nm - slit 8 nm).

Table 2. Reaction mixtures

Oxidase	Inducible acyl-CoA oxidase		Pristanoyl-CoA oxidase		THC-CoA oxidase	
	Blank	Test	Blank	Test	Blank	Test
KPi buffer	450	450	350	350	350	350
POD	180	180	140	140	140	140
BSA	22.5	22.5	-	-	70	70
Palmitoyl-CoA	-	225	-	-	-	-
2-CH$_3$-palmitoyl-CoA	-	-	-	175	-	-
THC-CoA			-	-	-	87.5
Water	1102.5	877.5	875	700	805	717.5
HVA	45	45	35	35	35	35
Dilution of Homogenates						
- Liver CT (A/D)	1/500	1/500	1/500	1/500	1/150	1/150
- Liver BF (B/E)	1/8000	1/8000	1/500	1/500	1/150	1/150
- Kidney CT (C/F)	1/200	1/200	1/100	1/100	1/50	1/50
Preincubation	25 mM NaN$_3$ 25 µM FAD		25 mM NaN$_3$ 25 µM FAD		25 mM NaN$_3$ 100 µM FAD	
Stop Reaction (min)	8 - 16 - 24 - 32		8 - 16 - 24 - 32		10 - 20 - 30 - 40	
Color Code	green		red		blue	

A,B,C = blank
D,E,F = test
CT = control
BF = bezafibrate

- Standardize fluorescence by means of uric acid and uricase as follows: Pipet in two sets of Eppendorf tubes 0 - 4 - 8 - 16 µl of uric acid; adjust to 16 µl with water; add to one set 80 µl of blank reaction mixture, to the other set test reaction mixture used to measure the inducible acyl-coA oxidase; start the reaction with 4 µl of diluted urate oxidase; place at 37°C; stop reaction after 20 min with 40 µl of HClO$_4$; proceed as described above for the homogenate samples.

1.3.4 Results and Comments

When palmitoyl-CoA was used as a substrate in the presence of albumin (molar ratio of plamitoyl-CoA to albumin = 10), an 8 to 13-fold increase in oxidase activity in liver was observed after bezafibrate treatment. Results were 2,228 ± 230 nmoles/min/g of liver for the control versus 23,769 ± 2,463 nmoles/min/g of liver after treatment (means ± S.E.M. of four experiments). The actual induction, however, must be greater since, under these assay conditions, approximately half of the measured palmitoyl-CoA oxidase activity is due to the action of pristanoyl-CoA oxidase, which is not affected by bezafibrate. Palmitoyl CoA oxidase activity in kidney of control rats was 440 ± 48 nmoles/min/g of kidney.

The activity of trihydroxycoprostanoyl-CoA oxidase, when expressed per g of liver, was not influenced by the proliferator (319 ± 48 and 342 ± 59 nmoles/min/g of liver in control and treated rats, respectively). In contrast to previous published data [7], we were able to detect in kidney of control rats some oxidase activity, using trihydroxycoprostanoyl-CoA as substrate (24.6 ± 2.7 nmoles/min/g kidney), presumably due to the higher sensitivity of the spectrofluorimeter used. This allowed us to use higher dilutions of the kidney homogenates, reducing the substrate-independent H$_2$O$_2$ formation in these assays, which interferes with the measurements.

With regard to the assay itself, the multiple manipulations serve specific purposes. Direct alkalinisation of the reaction mixtures causes a very rapid, time-dependent increase in the fluorescence values (irrespective of the presence or absence of substrate). The reason for this is not clear, but this increase is considerably less when homogenate proteins are first precipitated, followed by alkalinisation of the acidic supernatant. Denaturation by TCA should be avoided, since this compound causes quenching of the HVA-dimer fluorescence. Despite this extra step, one must still work on a time basis. Further investigations have shown that the addition of EDTA and dithiothreitol during alkalinisation suppresses the time-dependent increase in fluorescence almost completely. Both compounds are able to complex with divalent metal ions, suggesting that these ions (probably Fe^{2+}) are involved in the observed phenomena.

We prefer using the carbonate buffer since it gives the best result with regard to sensitivity and low background values. At the molarity used (0.5 M), a 15-fold dilution of the acidic supernatant results in the highest sensitivity.

Finally, we should mention that the HVA-POD assay can also be used in a kinetic mode, especially when measuring an oxidase which has a high activity or which tolerates the presence of detergents (e.g. the inducible acyl-CoA oxidase), since the HVA-dimer possesses some fluorescence at pH 7-8. However, the continuous illumination of the samples (excitation) leads to background problems, which are especially prominent when working with samples with a low oxidase activity (trihydroxycoprostanoyl-CoA oxidase in rat liver; L-pipecolic acid oxidase in human liver; acyl-CoA oxidases in fibroblasts).

References

1. Guilbault G, Kramer DN, Hackley E (1967) New substrate for fluorometric determination of oxidative enzymes. Anal Chem 39:271
2. Inestrosa NC, Bronfman M, Leighton F (1980) Purification of the peroxisomal fatty acyl-CoA oxidase from rat liver. Biochem Biophys Res Commun 95:7 - 12
3. Mannaerts GP, Van Veldhoven PP (1992) Role of peroxisomes in mammalian metabolism. Cell Biochem Funct 10:141 - 151
4. Osumi T, Hashimoto T, Ui N (1980) Purification and properties of acyl-CoA oxidase from rat liver. J Biochem 87:1735 - 1746
5. Schepers L, Van Veldhoven PP, Casteels M, Eysen HJ, Mannaerts GP (1990). Presence of three acyl-CoA oxidases in rat liver peroxisomes. An inducible fatty acyl-CoA oxidase, a non-inducible fatty acyl-CoA oxidase and a non-inducible trihydroxycoprostanoyl-CoA oxidase. J Biol Chem 265:5242 - 5246
6. Van den Bosch H, Schutgens RBH, Wanders RJA, Tager J (1992) Biochemistry of peroxisomes. Annu Rev Biochem 61:157 - 197
7. Van Veldhoven PP, Vanhove G, Vanhoutte F, Dacremont G, Parmentier G, Eysen HJ, Mannaerts GP (1991) Identification and purification of a branched chain fatty acyl-CoA oxidase. J Biol Chem 266:24676 - 24683
8. Van Veldhoven PP, Vanhove G, Asselberghs S, Eyssen HJ, Mannaerts GP (1992) Substrate specificities of rat liver peroxisomal acyl-CoA oxidases: Palmitoyl-CoA oxidase (inducible acyl-CoA oxidase), pristanoyl-CoA oxidase (non-inducible acyl-CoA oxidase) and trihydroxycoprostanoyl-CoA oxidase. J Biol Chem 267:20065 - 20074

2 Immunoblotting of Peroxisomal Proteins with Monospecific Antibodies

C. PACOT

2.1 Introduction and Aims

The most commonly used protein electrophoresis system today is the method described by Laemmli [2]. The Laemmli procedure, named sodium dodecylsulfate-polyacrylamide gel electrophoresis (SDS-PAGE), is a discontinuous system: the pretreated proteins are first concentrated in a stacking gel before entering the separating gel. Before electrophoresis, proteins are denatured using SDS, an anionic detergent, that wraps around the polypeptide backbone and confers a negative charge to the polypeptide in proportion to its length. Proteins are also treated with a reducing agent such as 2-mercaptoethanol, which breaks disulfide bonds. After treatment, polypeptides are separated on the support gel, a polyacrylamide matrix, on the basis of molecular weight by means of an electrical field in the presence of SDS. The major use of this system is to determine the molecular weight of polypeptides by running the gel silmutaneously with standard polypeptides of known molecular weights. A linear relationship exists between the $\log_{10}$ of the molecular weight of a polypeptide and its Rf (the distance from the limit between the stacking and separating gels to the polypeptide band, divided by the distance from the limit of the two gels to the front underlined by a low molecular weight dye). With this technique, we also gain information on the composition in subunits of a protein complex. After electrophoresis, the gel can be stained by several procedures: Coomassie blue stain is sensitive only to about 10 µg of a protein mixture, whereas silver stain is sensitive to about 10 ng; autoradiography can be used if labelled amino acids have been incorporated into the polypeptide. Once the gel is stained, it can be photographed or scanned by densitometry for a record of the position and intensity of each band.

A highly sensitive technique to characterize the presence, quantity or molecular weight of one or several specific proteins resolved by SDS-PAGE is immunoblotting. Proteins are transferred in an electric field from the slab gel to a membrane that binds the proteins non-specifically and can then be identified with specific antisera. Although there are a number of variations on the immunoblotting method [1], the most common and easiest technique is electro-transfer to nitrocellulose sheets [4]. Before immunodetection, the non-specific binding sites on the membrane are blocked to prevent background caused by further reactions with the membrane. Blocking can be achieved with a range of proteins (BSA, horse serum, non fat dry milk...) or detergent solutions (Tween 20, PBS...). The membrane is then incubated with a specific primary antibody. Three types of antibody preparations can be used for immunoblotting: monoclonal, polyclonal or pool of monoclonal antibodies. Polyclonal antibodies are probably the most widely used. As SDS denatures the structure of the proteins, only the antibodies that recognize epitopes resistant to the denaturation will bind. Most polyclonal sera contain at least some antibodies of this type, but only a few monoclonal antibodies will react with denatured antigens; thus polyclonal antibodies generate a relatively stronger signal than monoclonal antibodies. The major advantage of monoclonal antibodies is the high specificity of their interactions with the antigen. This is a good tool for identifiying a particular region of the antigen, with minimal background reactions as compared to polyclonal antibodies. However, as polyclonal sera contain high concentrations of specific antibodies, they can be diluted extensively to reduce background reactions without reducing sensitivity.

The primary antibody can be used labelled and can detect the location of the antigen directly, or it can be used unlabelled and located by a labelled secondary antibody (antiimmunoglobulin or protein A). A large number of methods exist to label the two antibodies, and which one(s) are used depends on the experiment. For example, we have iodine (^{125}I) labelling, in particular directly iodinated monoclonal antibodies, biotinylation of primary antibodies and enzymatic labelling of secondary antibodies, using enzymes such as peroxidase, alkaline phosphatase or β-galactosidase (for details see [1]). Whatever the choice for labelling, direct detection is less sensitive than indirect techniques and necessitates the purification and labelling of each antibody before use, although it results in a lower background effect and decreases the possibility of cross reactions. Most immunoblots should be analyzed using indirect methods. In our case, we will use the indirect method with an unlabelled antibody detected by a secondary antibody labelled with an anti-immunoglobulin coupled with alkaline phosphatase. For example, if the specific primary antibody is produced from rabbit serum, the second antibody will be an anti-rabbit IgG alkaline phosphatase conjugated antibody. Detection will be done with a bromochloroindolyl phosphate/nitro blue tetrazolium (BCIP/NBT) substrate, which generates an intense black/purple precipitate at the site of enzyme binding.

One example of electrophoresis and immunoblotting techniques will be done in this experiment by using rat liver peroxisomes and subfractions (see Experiment 1) and specific peroxisomal protein antibodies. Proteins from three different fractions, such as purified peroxisome fractions, peroxisomal membrane fractions and peroxisomal matrix fractions, will be separated on SDS-PAGE. After staining with Coomassie blue, the observation of the three different electrophoretic patterns could give some qualitative information on the relative content of several proteins and on the progressive purification of membrane and matrix subfractions prepared from purified peroxisomes.

The subperoxisomal localization of some peroxisomal proteins in the membrane (for example PMP 70) and the matrix (for example catalase and acyl CoA oxidase) can be determined by immunoblotting (for review, see [5]). In the present work, immunoblotting will be done with the available specific polyclonal antibodies anti-catalase and anti-acyl CoA oxidase (gifts from Völkl A., Heidelberg).

The fractions studied in this work will be isolated from liver of control rats and rats treated with fibrates. Fibrates are now well known as inducers of the expression of some peroxisomal protein genes [3]. We can therefore observe whether or not the proteins studied on the immunoblots are induced by fibrate-treatment.

2.2 Equipment, Chemicals and Solutions

2.2.1 Equipment

2.2.1.1 Access to

- a vertical slab gel electrophoresis apparatus for two simultaneous gels (Bioblock C98057) with a power supply.
- a semi-dry electrotransfer apparatus (Milliblot-SDE Transfert system, Millipore, MBB-DGE-002)
- a thermostated dry bath and a microcentrifuge for Eppendorf tubes
- a microwave oven
- a magnetic stirrer with teflon-coated magnetic rods
- a mixer and a 3D stirrer

2.2.1.2 On the Bench

- Pasteur pipettes
- adjustable automatic pipettes with tips
- a Hamilton syringe (100μl)
- Eppendorf tubes
- airtight plastic dishes for the incubation of transfer membranes
- a timer
- gloves
- Whatman 3MM paper
- transfer membrane, nitrocellulose (Hybond C, Amersham) or Immobilon P (Millipore IPVH 091 20)

2.2.2 Chemicals

- a kit of electrophoresis proteins: low molecular weight markers (LMW) from 14.4 to 94 kDa (Pharmacia 17-0446-01)

- bromophenol blue
- TEMED
- ammonium persulfate
- Coomassie brilliant blue R250
- Ponceau-S concentrate (Sigma P7767)
- naphtol blue black (Fluka 70490)
- non-fat milk powder
- bovin serum albumin (BSA) fraction V
- nitroblue tetrazolium and 5-bromo-3-chloro indoyl phosphate (kit Promega 53771)
- anti-rabbit IgG (whole molecule) alkaline phosphatase conjugate from goat (Sigma A8025)

2.2.3 Solutions

Ultra pure water (Milli Q, Millipore) is used to prepare all aqueous solutions and throughout the procedure.

2.2.3.1 *Separation of Proteins on SDS - PAGE (10%)*

- 1% agarose
- separating gel buffer (1.5 M Tris/HCl, 0.4% SDS, **pH 8.8**)
- stacking gel buffer (1.5 M Tris/HCl, 0.4% SDS, **pH 6.8**)
- 30% acrylamide, 0.8% bisacrylamide
- 4 x loading buffer (8% SDS, 40% glycerol, 0.1 M 2-mercaptoethanol)
- 0.05% bromophenol blue
- running buffer (28.8 g glycine, 6 g Tris, 1 g SDS, H_2O up to 1 L)
- Coomassie staining solution (0.25% Coomassie brilliant blue, 45% methanol, 9% acetic acid)
- gel destaining solution (30% ethanol, 10% acetic acid)
- gel drying solution (10% ethanol, 1% glycerol)

2.2.3.2 *Electrotransfer of Proteins on Nitrocellulose Membrane or Immobilon P Using a Semi Dry Apparatus*

- transfer buffer (25 mM Tris, 192 mM glycine, 15% methanol, 0.1% SDS)
- Ponceau-S working solution: 1 bottle (20 ml) of Ponceau-S concentrate + 180 ml water
- amido black staining solution (0.2% naphtol blue black, 7% acetic acid)
- amido black destaining solution (30% methanol, 10% acetic acid)

2.2.3.3 Blotting of Membrane with Specific Antibodies

- 10 x TBS buffer (100mM Tris/HCl pH8, 1.5 M NaCl, **pH8**)
- TBST buffer (0.05% Tween 20 in 1x TBS)
- 1 % BSA in TBST buffer
- 0.5 M NaCl in TBST buffer
- alkaline phosphatase buffer (AP buffer; 100 mM Tris/HCl, 100 mM NaCl, 5 mM $MgCl_2$, **pH9.5**)

2.3 Experimental Procedure

2.3.1 Electrophoresis of Proteins

2.3.1.1 Preparation of the Mini-Gels

For separation of peroxisomal proteins, SDS - 10% PAGE gels are used. One gel will be used for Coomassie blue staining and the other one for immunoblotting.

Note: The following volumes are given for 2 gels of 6 x 8 cm and 0.75 mm thick.

Separating Gel (10%):
- 2.25 ml separating gel buffer
- 3.75 ml H_2O
- 3 ml of 30% acrylamide-0.8% bisacrylamide solution
- 15 μl TEMED
- 30 μl 10% APS (good for 1 week)

Mix well, pour between the prepared plates with a Pasteur pipette and gently layer 1 mm of butanol on top to form a flat surface while acrylamide polymerizes. Wait for polymerization for about 30 min. After polymerization is complete, discard butanol, rinse with H_2O several times and dry with a filter paper.

Note: During this time, start to prepare the samples (see 2.3.1.2)

Stacking Gel (4%):
- 0.75 ml stacking gel buffer
- 1.85 ml H_2O
- 0.4 ml of 30% acrylamide-0.8% bisacrylamide solution
- 7.5 μl TEMED
- 15 μl 10% APS

Mix well, pour on the top of the separating gel and immediately insert the sample position forming comb (for a 0.75 mm-thick comb with 10 wells, maximum volume to load is 25 μl/well). Wait for polymerization for about 30 min. Remove the comb and wash the wells with H_2O (add and remove water to and from each well a few times), using a Hamilton syringe. Fit the gel in the electrophoresis unit and add running buffer.

2.3.1.2 Preparation of the Samples

Prepare the different samples in Eppendorf tubes as follows:

- peroxisomal proteins (10 - 20 μl). See results for an example of quantities of proteins loaded on the gel. If the volume of proteins is less than 10 μl, add H_2O up to 10 μl to have a sufficient volume to load.
- 6 μl of 4 x loading buffer

Mix well, centrifuge the tubes for a few seconds (to collect all of the sample in the bottom of the tubes), and heat to 95°C for 3 minutes. Add 2 μl of 0.05% bromophenol blue and centrifuge the tubes one more time before adding the samples into the wells with a Hamilton syringe. Don't load any sample in the two wells on the edges of the gel. Fill the empty wells with loading buffer.

Note: Don't forget to load molecular weight markers.

Dissolve the lyophilized mixture Pharmacia of 6 markers (100 μg of each) in 100μl of 2.5% SDS, 5% 2-mercaptoethanol. Heat an aliquot (4 μl of total marker proteins for the gel to be stained with Coomassie Blue and 10 μl for the gel to be blotted) at 95°C for 3 min and add 2 μl of bromophenol blue.

2.3.1.3 Electrophoresis Running

Run the gels at 20 mA for about 1 h to make proteins concentrate at the top of the separating gel. Run the gels at 40 mA for about 30 min to make proteins migrate in the separating gel.

2.3.1.4 Staining a Gel with Coomassie Brillant Blue

Remove the gels from the plates. Discard the stacking gel, and incubate one of the saparating gels in Coomassie staining solution for about 1 h with constant agitation. Destain the gel by washing with agitation several times. Incubate in drying solution. The gel can be dried quickly with a gel drier or slowly between 2 sheets of cellophane on a glass plate.

2.3.2 Electrotransfer of Proteins Using a Semi-dry Apparatus

Cut Whatman paper and membrane to the size of the gel. Equilibrate the gel, the Whatman paper and the nitrocellulose membrane in transfer buffer for 15 min .

Note: Generally, nitrocellulose membranes must be pre-equilibrated in several solutions, as described in the manual.

In case of using Immobilon P, wet the membrane in methanol for 1 to 3 seconds and immerse the membrane for 1 to 2 min in H_2O.

After equilibration, assemble the sandwich as follows:

TOP ·

———————————	1 (or more) sheets of Whatman filter
———————————	paper
▨▨▨▨▨▨▨▨▨▨▨▨	gel
▰▰▰▰▰▰▰▰▰▰▰▰	membrane
———————————	1 (or more) sheets of Whatman filter
———————————	paper

BOTTOM+

Keep the sandwich wet and avoid air bubbles by rolling a glass pipette over the surface. Connect the semi-dry apparatus to a power supply and run as suggested by the apparatus manual (100mA for 30 min with a gel of $40cm^2$). After transfer, the gel can be stained by Coomassie brillant blue and destained as usual to check that the transfer is complete.

Blotted proteins can be stained with different methods.In our case, the membrane is incubated in Ponceau-S working solution for 10 min and rinsed with H_2O several times. You can mark the position of molecular marker bands with a pencil. Ponceau-S staining is not very sensitive and fades very quickly after drying the membrane. But this staining has the advantage of being reversible, so the same blot can be used for probing with specific antibodies.

Molecular weight markers can be also stained with amido black staining solution. Cut the piece of membrane with the molecular weight markers and float it in the amido black solution briefly until bands become visible. Rinse several times, first with destaining buffer and then with water. The membrane can then be air dried.

2.3.3 Blotting of the Membrane with a Specific Antibody

All incubations are made at room temperature with gentle agitation in a small enough plastic dish to minimize the volume of solutions used.

2.3.3.1 Blocking Step

Incubate the membrane with TBST + 5% non-fat milk powder for 60 min (25 ml/100 mm^2 of plastic dish).

2.3.3.2 Primary Antibody Binding Step

Dilute the specific antibody (anti-acyl CoA oxidase type I protein or anti-catalase) with TBST + 1% BSA. Incubate the membrane with the diluted antibody for 60 min (10 ml/100 mm^2 of plastic dish).

2.3.3.3 Washing Steps

Wash the membrane 3 times with TBST + 0.5 M NaCl (each time for 10 min).

2.3.3.4 Second Antibody-Alkaline Phosphatase Conjugate Binding Step

Dilute the anti-rabbit IgG-alkaline phosphatase conjugate with TBST to 1:1000 dilution. Incubate the membrane with the diluted conjugate for 30 min (10 ml/mm^2 of plastic dish).

2.3.3.5 Washing Steps

Wash the membrane 3 times with TBST + 0.5 M NaCl (each time for 10 min).

2.3.3.6 Alkaline Phosphatase Reaction Step

Prepare the color substrate solution just before use: mix 6.6 µl nitroblue tetrazolium substrate/ml AP buffer, then add 3.3 µl 5-bromo-3-chloro-indoyl phosphate/ml AP buffer and mix. Incubate the membrane with color substrate solution (10 ml/100 mm^2 of plastic dish) and time the incubation to observe the development (30 sec to 5 min). Stop the reaction by washing the membrane with H_2O several times. The membrane can be stored dry.

2.4 Results

Figure 1: The gel stained with Coomassie blue shows 2 major bands at 66 and 34 kDa in the control peroxisome fraction, which could correspond to catalase and urate oxidase, respectively . Ciprofibrate treatment induces a strong increase for the 75 kDa band in both the homogenate and peroxisome fractions. Several proteins, such as enoyl CoA hydratase, carnitine octanoyl transferase and acyl CoA synthetase, have a molecular weight at about 75-80 kDa. Other bands are also increased in the treated rat peroxisome fraction: at 72 and 50 kDa, likely the 2 subunits A and B of acyl CoA oxidase, and at 40 kDa, likely thiolase. The third subunit AOX C, at about 22 kDa, is difficult to observe on this gel. We can see a relative decrease of the catalase band at 66 kDa. This result doesn't reflect the 2 fold increase in the enzyme activity and in the mRNA amount generally observed after a treatment with ciprofibrate. The decrease in the intensity of the band is due to a decrease in the percentage of catalase in total peroxisomal proteins from treated rats in comparison with control rats. Similar explanation for the relative decrease of the urate oxidase band can be put forward.

Figure 2A: immunoblot with anti-acyl CoA oxidase antibody. Ciprofibrate treatment induces the three subunits of acyl CoA oxidase at 72, 50 and 21 kDa in the peroxisome and homogenate fractions. The corresponding bands appear to be absent in the homogenate of control rats because of the very low amount of acyl CoA oxidase in this fraction.

Figure 2B: immunoblot with anti-catalase antibody. Concerning the peroxisome fraction, the relative decrease in the intensity of the catalase band in treated rats as compared to control rats is explained in Figure 1. In the homogenate fractions, catalase represents a low amount of total proteins. However, we can observe a slight increase in the catalase band after a treatment with ciprofibrate.

Exp. 2 - Immunoblotting of peroxisomal proteins

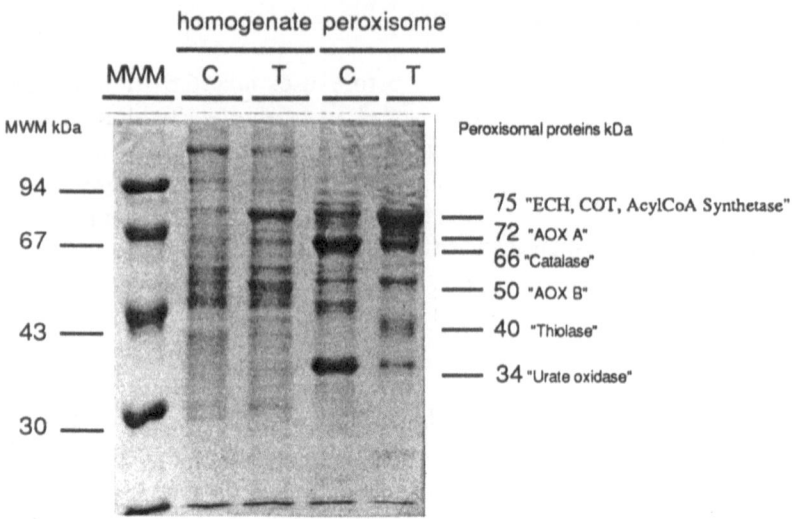

Figure 1: SDS-10 % PAGE of liver homogenate and purified peroxisome proteins from rats. 10 µg of homopgenate and peroxisome proteins, and 4 µl of MWM mixture were loaded on the gel, electrophoretised and stained using Coomassie blue.

MWM: molecular weight marker
AOX: acyl CoA oxidase
ECH: enoyl CoA hydratase
COT: carnitine octanoyl transferase
C: control rats
T: rats treated by oral administration of ciprofibrate at 3 mg/kg/d during 2 weeks

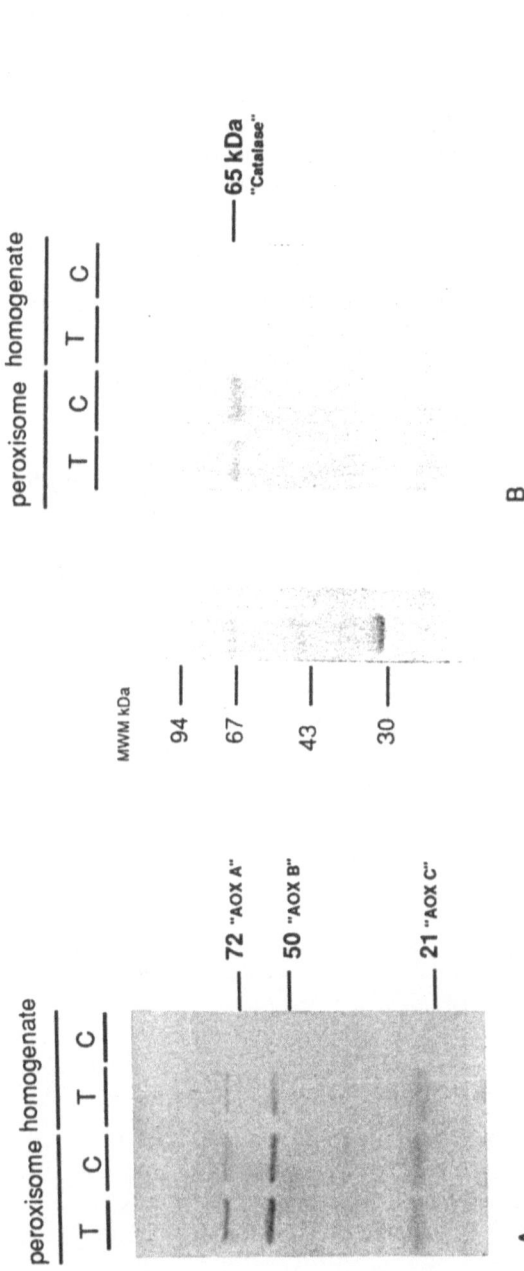

Figure 2: Immunoblots of homogenate and purified peroxisome proteins from rat liver, performed with anti-acyl CoA oxidase (A) or anti-catalase (B) antibody and detected with the phosphatase alkaline method. 2 μg of homogenate and peroxisome proteins were loaded on the gel for immunoblotting. Membranes were incubated with the first antibody diluted to 1/200 during 1 h and then with the second alkakine phosphatase-conjugate antibody (see experimental procedure) diluted to 1/1000 during 35 min. The piece of membrane with MWM was stained with amido black.
MWM: molecular weight marker; *AOX:* acyl CoA oxidase
C: control rats; *T:* rats treated by oral administration of ciprofibrate at 3 mg/kg/d during 2 weeks

References

1. Harlow E, Lane D (1988) In "Antibodies, A laboratory manual.", Cold Spring Harbor Laboratory, New York, USA
2. Laemmli, EK (1970) Cleavage of structural proteins during the assembly of the head of bacteriophage T4. Nature 227:680-685
3. Nemali MR, Usuda N, Reddy MK, Oyasu K, Hashimoto T et al. (1988) Comparison of constitutive and inducible levels expression of peroxisomal ß-oxidation and catalase genes in liver and extrahepatic tissues of rat. Cancer Res 48:5316-5324
4. Towbin H, Staehelin T, Gordon J (1979) Electrophoretic transfer of proteins from polyacrylamide gels to nitrocellulose sheets: Procedure and some applications. Proc Natl Sci USA 76:4350-4354
5. Van den Bosch H, Schutgens RBH, Wanders RJA, Tager JM (1992) Biochemistry of peroxisomes. Annu Rev Biochem 61:157-197

Acknowledgements

The organizers wish to thank Dr M. Petit (Sterling Winthrop, Dijon) for his help in animal treatment during the FEBS practical course.

3 Morphology of Peroxisomes in Light- and Electron Microscopy

E. BAUMGART

3.1 General Introduction

Peroxisomes (POs) vary markedly in distribution, morphological shape and enzyme composition in different organs and cell types (for a review see [14]). In the original studies of DE DUVE and associates (for a review see [9]), POs were considered to be present only in a few cell types, like hepatocytes and renal tubular epithelial cells. The development of the alkaline 3,3'-diaminobenzidine (DAB) method for the light- and electron microscopical visualization of the peroxidatic activity of catalase (a marker enzyme of POs)[10, 11] led to the discovery of the ubiquity of this cell organelle [18]. Since POs are involved in the metabolism of long-chain fatty acids, they are very numerous and large in organs metabolizing fatty acids like the liver. Hepatic POs are spherical cell organelles with a single limiting membrane surrounding a homogeneous, finely granular matrix. In some mammalian species (like beef, sheep, cat, rodents), hepatic, or even renal, POs may contain crystalline inclusions in their matrix (cores = crystallized urate oxidase [1, 25]; marginal plates = crystallized α-hydroxy acid oxidase B [29, 30].

POs are cell organelles with a high plasticity undergoing proliferation and exhibiting alterations of their morphological appearance and enzyme composition, induced by different stimuli (like changes in lipid metabolism, treatment with hypolipidemic drugs, different thyroid states, regenerating rat liver [3, 4, 5, 15, 17, 21, 27, 28]; for reviews see [14, 22, 23]). Since PO-proliferation is not necessarily associated with significant alterations of the marker enzymes of this cell organelle [5], morphological investigations in combination with biochemical experiments are necessary to characterize the exact nature of the peroxisomal response.

3.2 Overall Aims

Fixation and processing of tissue or cells may vary markedly for the cytochemical demonstration of the activity of peroxisomal enzymes and for the localization of the peroxisomal proteins by immunocytochemistry (for a review see [12]).

The following is a short overview of the four major morphological methods for the demonstration of POs:

1) Indirect immunofluorescence with fluorescein isothiocyanate (FITC)- or tetramethylrhodamine isothiocyanate (TRITC)-labeled secondary antibodies:
This method may be applied for the light microscopical visualization of POs in cell culture (i.e. hepatoma cell cultures or cultures of fibroblasts from patients with peroxisomal diseases), or on frozen sections of various tissues.

2) Immunohistochemical detection of peroxisomal antigens with the post-embedding protein A-gold-silver intensification-technique:
This method may be applied to paraffin sections [19] and to unosmicated Epon 812- and LR White-semithin sections [20, 25]. It is particularly useful for the demonstration of changes in the distribution or the number of the POs using automatic image analysis [6].

3) Immunocytochemical localization of peroxisomal enzymes with the post-embedding protein A-gold technique:
This method may be applied for the demonstration of peroxisomal proteins for electron microscopy [4, 8, 25].

4) Fixation, processing and embedding of peroxisomal fractions for routine electron microscopy [26], as well as for the cytochemical localization of the catalase activity with the alkaline DAB-method [10, 11].

3.3 Indirect Immunofluorescence with FITC- or TRITC-Labeled Secondary Antibodies

3.3.1 Aim

Double-localization of catalase and 3-enoyl-CoA hydratase/dehydrogenase in peroxisomes of Hep G2 cells

3.3.2 Equipment

3.3.2.1 Access to

- a fluorescence microscope (Olympus BHS) equipped with 63 X (or 100 X) phase contrast objectives (oil immersion) and FITC- and TRITC- absorption/emission filters
- Kodak TMY400-films or ASA 400 color slide films
- a balance
- 4°C and -20°C refrigerators
- a cell culture room with a CO_2-incubator, a clean bench with a gas burner and a water bath

- minimal essential medium (α-MEM) with 10% foetal calf serum (FCS) and penicillin/streptomycin
- sterile cell culture pipettes (2 ml and 10 ml)
- an autoclave
- a hood for chemical vapors

3.3.2.2 On the Bench

Basic Equipment
- a pH-meter
- a magnetic stirrer with a heating system and magnetic rods
- an Eppendorf microcentrifuge
- a vortex mixer
- timers
- a crushed ice container
- glassware
- sterile and non-sterile Pasteur pipettes and a pipetting aid device
- adjustable automatic pipettes and sterile and non-sterile tips
- Eppendorf tubes (sterile and non-sterile) and racks
- aluminium foil
- a plastic squeeze bottle for distilled water
- Parafilm
- paper towels
- very fine forceps (Dumont tweezers n°7, Ladd Research Industries)
- Whatman filter paper
- lense paper
- gloves

Additional Equipment
- a tap-water vacuum pump with a vacuum bottle for collecting the washing buffers
- round glass cover slips (0.8-1.5 cm)
- clean glass slides for fluorescence microscopy
- multiwell plates with 6 or 12 big wells (bigger than 1.5 cm) or sterile Petri dishes (3.5 cm), used for the preparation of a moist chamber

3.3.3 Cells, Chemicals and Solutions

- Hep G2 cells (available from ATCC = American Type Culture Collection) cultured for two days on glass cover slips

- 1 x calcium and magnesium free phosphate buffered saline (1 x CMF-PBS): 0.14M NaCl, 2.7mM KCl, 8.1mM Na_2HPO_4, 1.5mM KH_2PO_4, pH 7.4. CMF-PBS should be prepared as a 10 times concentrated stock solution (per liter use: 80g NaCl, 2g KCl, 14.4g Na_2HPO_4, 2 H_2O and 2g KH_2PO_4)
- 4% p-formaldehyde in 1 x CMF-PBS, pH 7.4
- 0.2% Triton X-100 in 1 x CMF-PBS
- 1% glycin in 1 x CMF-PBS
- specific antibodies: anti-catalase (mouse-polyclonal)- or anti-hydratase (rabbit-polyclonal)-antibodies.
- fluorochrome-labeled secondary antibodies (all purchased from Sigma GmbH, Deisenhofen, FRG):

  - 1) goat-anti-rabbit IgG-FITC-labeled
  - 2) goat-anti-mouse IgG-TRITC-labeled
     or
  - 3) goat-anti-rabbitIgG-TRITC-labeled
  - 4) goat-anti-mouse IgG-FITC-labeled

- Mowiol 4.88 (Merck) in glycerol containing n-propyl gallate (Sigma)
- 70% ethanol
- ultrapure water
- crushed ice

3.3.4 Experimental Procedure

- Prepare the antibody dilutions for the first antibodies (usually 1:200 - 1:300 if the IgG-fractions contain 10 µg/µl protein according to Lowry's procedure) before the experiment or during the fixation time; keep on ice prior to use.
- Prepare the fixative (p-formaldehyde must be heated to depolymerize under a hood; add NaOH to dissolve the p-formaldehyde). Cool to room temperature.
- Take out the cells from the CO_2-incubator, and remove the culture medium with a Pasteur pipette connected to the vacuum bottle and vacuum pump.
- Wash the cells on the cover slips with ca. 1.5ml 1 x CMF-PBS.
- Carry out all washing steps by pipetting the solutions carefully into the Petri dishes (or multiwells), gently shaking and removing the solutions with the vacuum pump.
- Make sure that the cells never dry during the whole incubation procedure.
- Fix the cells immediately for 20 min at room temperature with 4%-p-formaldehyde in 1 x CMF-PBS.
- Wash 3 times with 1.5ml CMF-PBS.
- Permeabilize the cells with 0.2% Triton X-100 in CMF-PBS (1-2ml per dish) for 15 min at room temperature, not longer!
- Wash 3 times with 1.5ml 1 x CMF-PBS.

- Block free aldehyde groups with 1.5ml 1% glycine in 1 x CMF-PBS for 10 min at room temperature.
- In the meantime prepare a moist chamber using a new Petri dish and wet Whatman filters.
- Wash the cells 3 times with 1.5ml CMF-PBS.
- Take out the cover slips, dry briefly the side without cells and insert into the moist chamber (cell side up!).
- Apply carefully (without making bubbles) 20 µl of diluted antibody solution onto the middle of the cover slips; be sure that the solution will not spread onto the surrounding Petri dish
- Incubate overnight at 4°C in the refrigerator or at room temperature (if overnight-incubation is not possible, incubate at 37°C at least for 30 min); overnight-incubation will yield better immunoreactivity; if the incubation is done at room temperature, you should make sure that the cells with antibody solutions will not dry!

The Next Morning:

- Wash the cover slips 3 times with 1.5 ml 1 x CMF-PBS.
- Prepare a new moist chamber with dry bottom.
- Place the cover slips (cell side up, bottom dry) into the new moist chamber.
- Apply 20µl of secondary antibody solution (usually diluted 1:100 to 1:200 in 1 x CMF-PBS), and incubate at least for 30 min at 37°C.
- During the incubation time prepare glass slides; clean with 70% ethanol, dry and clean with lense paper; mark each slide (one slide each for one cover slip) with a waterproof marker, and add a small drop of Mowiol 4.88 in glycerol (without making air bubbles); store the slide dust-free until the end of the incubation procedure.
- After incubation with the secondary antibodies, wash the cover slips again 3 times with 1.5ml 1 x CMF-PBS.
- Take out each cover slip with fine forceps (be very careful since the cover slips break very easily).
- Dip the cover slips briefly into distilled water and remove the water by holding a piece of Whatman filter close to the rim of the cover slip.
- Place the cover slip (cell-side down, facing the glass slide) onto the drop of the Mowiol 4.88 on the glass slides; be careful not to enclose air-bubbles.

The cells are now ready to be examined without oil immersion under the fluorescence microscope. Don't use fresh preparations for oil immersion, since the Mowiol 4.88 does not polymerize very fast.

Since POs of cultured cells are very small it is necessary to use 63X or 100X oil immersion objectives to clearly visualize the small organelles. Therefore, store the slides overnight in complete darkness; you can use the slides the next morning for fluorescence microscopy at higher magnifications.

Document the results using Kodak TMY 400 film for black and white paper prints or a ASA 400 film for color slides.

The exposure time depends on the amount of fluorescence, but should usually not exceed 1 min.

3.3.5 Expected Results

The Hep G2-cells contain round or elongated POs that are clustered around the nuclear region; only very few POs are found in the periphery of the cells.

If FITC-labeled antibodies are used for the detection, the POs will be seen using the FITC-filter as small green dots, scattered around the nucleus (dark region); if TRITC-labeled antibodies and -filters are used the POs should appear in dark red.

In double-labeling experiments with catalase- and hydratase-antibodies, the POs should be superimposable using FITC-and TRITC-filters.

3.3.6 Trouble Shooting

High cytoplasmic background staining: the concentration of the antibodies were too high; test different first- and second-antibody concentrations

No immunological staining at all: the following reasons may cause this problem:

- 1) Cell permeabilization may be insufficient; 0.05% Tween 20 may be included in antibody incubation- and washing solutions.
- 2) Antibody dilutions may be either too high or too low.
- 3) The specific antibody may have lost its binding capacity; check in another system, such as immunoblots.

3.4 Immunohistochemistry (1 μm LR White-sections) with the Post-Embedding-Protein A-Gold-Silver Intensification-Technique

3.4.1 Aim

Immunohistochemical localization of catalase in POs of normal and bezafibrate-treated rat liver; demonstration of peroxisomal proliferation.

3.4.2 Equipment

3.4.2.1 Access to

- an ultramicrotome (Reichert Ultracut S) for the demonstration of the sectioning technique
- a knife maker (Reichert Knifemaker II) and glass stripes for making glass knifes
- a fluorescence microscope (Olympus BHS) equipped with epipolization filters or a regular light microscope
- 4°C and -20°C refrigerators
- a precision balance

3.4.2.2 On the Bench

Basic Equipment (see 3.3.2.2)

Additional Equipment
- a 37°C water bath shaker (must be light-tight for the silver intensification reaction!)
- a clean glass plate (ca. 10 x 20 cm)
- a plastic cover that fits exactly on the glass plate
- big plastic forceps (ca. 15 cm length)
- Falcon tubes (50ml) with a rack
- 4 glass cuvettes (50ml) for glass slides as used in staining procedures in histology (for example Coplin cuvettes)
- a light-tight box (ca. 10 x 10 x 6 cm)
- 10ml plastic incubation vials with tops
- a diamond pencil (for marking the glass slides)
- cover slips (ca. 18 x 18mm)

3.4.3 Sections, Chemicals and Solutions

- Precut semithin sections (1μm) of liver tissue (embedded for immunocytochemistry in LR White) on silanized glass slides; Never use egg-white for adhesion of the sections, as this will yield precipitation of silver onto the sections!
- Use Tris-buffered saline-albumine (TBSA: 20mM Tris, 0.9% NaCl and 0.1% bovine serum albumin, pH 7.4) for the incubation- and washing-buffers.
- Prepare also a small amount of TBS without BSA. This solution will be needed for preparing the fixative (2% glutaraldehyde in TBS)
- glutaraldehyde (ultrapure, 25% aqueous stock solution, Serva)
- 1% BSA in TBS

- primary antibodies (from Prof. Völkl, Heidelberg): a monospecific polyclonal rabbit antibody against rat liver catalase, IgG-fraction with 10μg/μl protein content (usually diluted 1:250 to 1:1000 in TBSA for the incubation)
- protein A-gold solution, prepared according to the protocol of [24]; (gold particle size: 6 or 15 nm; extinction 0.3 to 0.5; diluted 1:50 to 1:75 in TBSA for the incubation)
- ultra pure water
- silver lactate (Sigma)
- hydroquinone
- 0.5M Na-citrate buffer, pH 4.0
- arabic gum (25g in 50 ml ultrapure water; must be stirred overnight and filtered through cotton-cloth prior to use)
- photographic fixing solution (Allfix, Fesago), diluted 1:10 for use
- DePex (Serva)

3.4.4 Experimental Procedure

- Cut 1 μm LR White sections (with an ultramicrotome) onto presilanized glass slides; dry and store the slides in a slide box prior to further use.
- Mark the place of the section using a diamond pencil.
- Place the glass slides onto the glass plate.
- Block the non-specific binding sites by adding ca. 100μl of the 1% BSA-solution on top of the section (spread the drop to the marked scratch, and be careful not to make air-bubbles); incubate for at least 30 min at room temperature.
- Cover the glass slides at each incubation step with the plastic cover to prevent dust deposition onto the sections.
- During the blocking-time prepare the dilutions of the primary antibodies (usually 1:250 to 1:1000 in TBSA); store on ice.
- Remove the blocking solution with a Pasteur pipette connected to the vacuum bottle and vacuum pump.
- Add ca. 50-100μl of the primary antibody dilution for each section; incubate overnight at room temperature; to prevent the sections from drying, wet four pieces of Whatman filters, roll and place them on each side of the section onto the glass plate. Close the cover tightly!

The Next Morning:

Protein A-Gold Fixation
- Remove the antibody solution with the Pasteur pipette connected to the vacuum bottle.
- Wash the sections on the glass slides 5 times for 3 min. each with TBSA by

pipetting ca. 500μl to 1000μl onto each section and then removing TBSA with the vacuum pump.
- After removing of the washing solution, add 50 to 100μl of protein A-gold solution, diluted 1:50 to 1:75 in TBSA; incubate for 90 min at room temperature, not longer!
- During the incubation time, preheat the light-tight water bath shaker to 37°C.
- Wash the sections 5 times for 3 min each with 500-1000μl TBSA.
- Fix the antibody-protein A-gold-complex for 15 min with 2% glutaraldehyde in TBS without BSA at room temperature; always freshly prepare the fixative before the silver intensification.
- Wash the slides again 3 times with TBS.
- Rinse the slides with distilled water and air dry (dustfree!).
- Now the slides are ready for the silver intensification.

Silver Intensification
- Dissolve 0.43g hydroquinone in 7.5ml distilled water in 10ml plastic vials (hydroquinone dissolves very slowly; put the solution into the waterbath shaker at 37°C in darkness and shake)
- Prepare 55 mg of silver lactate in a 10ml plastic vial and store in the light-tight box at room temperature until use.
- Prepare the following mixture in a Falcon tube (50ml) (one Falcon tube for 2 slides): 30 ml arabic gum solution, 5 ml 0.5M Na-citrate buffer pH4.0, and 7.5 ml hydroquinone
- Heat the solution to 37°C in the waterbath shaker.
- Prepare two 50ml glass slide cuvettes: one with Allfix (1:10 in distilled water), the other one with distilled water.
- Preincubate two glass slides with the sections (back to back) in the Falcon tube for 5 min in darkness at 37°C with gentle rocking; never touch the glass slides with bare hands; always use plastic forceps and gloves!
- In the meantime, dissolve the preweighed silver lactate in 7.5 ml distilled water (keep in darkness).
- For silver intensification of the sections, quickly open the Falcon tubes and pour the silver lactate solution into the tubes, close them and mix rapidly by repeated inversion of each tube. Close the waterbath shaker and incubate for 7-10 min at 37°C in complete darkness
- Take out the slides with plastic forceps and rinse them under running distilled water.
- Fix the slides exactly for 2 min in the glass slide cuvette with 50 ml Allfix (1:10 diluted in distilled water) at room temperature.
- Rinse the slides after fixation in a new 50 ml cuvette containing distilled water; wash 3 times for 5 min each with shaking.
- Air dry the slides.
- Clean cover slips with lense paper.
- Enclose the sections with DePex.

The sections are now ready for normal light or epipolarization microscopy. Be careful, as the enclosing substance may not be polymerized. The aggregates of gold-silver-particles will fluoresce to a yellowish tinge in epipolarization microscopy; in normal light microscopy, you will observe the silver aggregates as black dots.

3.4.5 Expected Results

- There is always a small rim of silver particles around the sections (due to BSA-binding to silanized slides.
- The POs are visible as small dots (black in normal light, yellowish-fluorescing in epipolarization microscopy).
- Hepatic sinusoids are completely negative!
- In normal rat liver, POs are randomly distributed in the cytoplasm of hepatocytes, and no lobular gradient is visible.
- In bezafibrate-treated rat liver, POs are proliferated and usually form aggregates of 5-10 POs.
- With some other hypolipidemic drugs (like BM 15766 or gemfibrozil; [3, 16]), a lobular gradient, with stronger proliferation of POs in the perivenous region, will be visible.
- The proliferation of POs can be assessed now with an automatic image analysis system (TAS, Leitz, Wetzlar; [6]).

3.4.6 Trouble Shooting

Precipitates of silver in sinusoids of the liver: several reasons may cause this problem:

- 1) Blocking may be insufficient; enhance the concentration of BSA in the blocking solution or increase the blocking time.
- 2) The concentration of the specific antibody may be too high.
- 3) The concentration of the protein A-gold solution may be too high.
- 4) The washing steps may not have been carried out carefully; increase the amount of washing buffer and wash more intensively.
- 5) The silver intensification time may be too long, or the reaction has not been carried out in complete darkness

Negative or weak silver precipitates:

- If no silver ring is observed around the sections, the silver intensification solution was not prepared correctly. Wash the slides carefully with water, fix again with 2% glutaraldehyde-TBS, and try again. Glutaraldehyde-fixation is necessary for the amplification reaction.

- If the silver ring is visible, but there is no reaction product in the hepatocytes, check the antibody- or protein A-gold solution. Use protein A-gold with smaller gold particle size for the detection, since smaller particles will bind much better to the Fc-fragment of the antibody and will yield a stronger signal after silver intensification.
- If the signal is only very faint, try longer silver intensification times.

3.5 Immunocytochemistry with the Postembedding Protein A-Gold Technique

3.5.1 Aim

Electron microscopical demonstration of peroxisomal subcompartmentation by localizing catalase and urate oxidase antigenic sites

3.5.2 Equipment

3.5.2.1 Access to

- an electron microscope (Hitachi H600)
- an ultramicrotome (Reichert Ultracut S) and knife maker (Reichert Knifemaker II) with glass stripes for the demonstration of the sectioning technique
- 4°C and -20°C refrigerators
- a balance

3.5.2.2 On the Bench

Basic Equipment (see 3.3.2.2)

Additional Equipment
- a timer indicating seconds (for contrasting)
- a grid box
- Petri dishes (3.5cm and 9cm) with round filter paper fitting into them
- a glass or plastic cover (ca. 10 x 20 cm)

3.5.3 Sections, Chemicals and Solutions

- LR White sections (ca. 70nm thickness) of normal rat liver tissue, cut on formvar-coated nickle-grids (200 mesh)

47

- Tris buffered saline-albumine (TBSA: 20mM Tris, 0.9% NaCl and 0.1% bovine serum albumine)
- 1% BSA-TBS solution for blocking
- specific antibodies (provided by Prof. Völkl, Heidelberg), anti-rat-catalase and urate oxidase, monospecific polyclonal rabbit IgG-fractions (10μg/μl protein according to Lowry's procedure); for incubation, dilute them 1:500 to 1:5000.
- protein A-gold solution, prepared as described by [24], E = 0.53, 6 or 15 nm gold particle size, diluted 1:50 to 1:75 prior to use
- sterile water (aqua ad injectabilia, in glass bottles, as used for injections in hospital)
- 2% uranyl acetate solution (aqueous)
- Reynolds lead citrate for contrasting sections

3.5.4 Experimental Procedure

- Carry out all incubation and washing steps at room temperature; never let the grids with sections dry during incubation; centrifuge all incubation solutions (especially protein A-gold) for 2 min in an Eppendorf microcentrifuge prior to use.
- Put a small piece of Parafilm into a small Petri dish (clean side up; don't touch with fingers) and place as many 50μl drops of 1% BSA-TBS solution onto the film as grids are to be incubated.
- Carefully take the grids out of the grid box with fine forceps (touch the grid only at the rim) and place each grid onto a drop of 1% BSA-TBS (section side down).
- Block the nonspecific binding sites for at least 30 min and, in the meantime, prepare the antibody-dilutions (1:500 to 1:5000 in TBSA).
- Prepare a moist chamber for overnight incubation as follows: take a big Petri dish, insert several wet filter papers, and place a small Petri dish onto the wet filters inside the larger one; mark the position of different antibody drops and grid numbers, overlay the markation with clean Parafilm and spot 50μl of the appropriate antibody solutions onto the marked places.
- After blocking, take the grids carefully off of the BSA-drops and transfer them without drying or other manipulations to a drop of the appropriate antibody, section side down.
- Incubate overnight at room temperature.

The Next Morning:

- Wash the grids by transferring them to a series of at least 10 drops (each 50μl) of TBSA. For each grid, place the series of drops onto clean Parafilm and mark the appropriate line with the grid number; move the grids from one drop to the

next, let stand for 1 or 2 min and go on washing.
- Always cover the drops with a plastic or glass cover; otherwise many small dust particles will be attracted to the grids.
- Prepare the protein A-gold solution (diluted 1:50 to 1:75 in TBSA); don't forget to centrifuge before use; otherwise a lot of gold aggregates may precipitate onto the sections; incubate for 90 min, but not longer!
- Wash the grids under a stream of ultrapure water using a plastic washing bottle.
- Air dry.
- Contrast the sections for 2 min on drops of uranyl acetate (2% in ultrapure water).
- Wash briefly under a stream of water.
- Contrast for 45s to 1 min on a drop of lead citrate.
- Wash extensively under a stream of water.

The sections are now ready to be examined in an electron microscope.

3.5.5 Expected Results

The POs should be heavily labelled with gold particles, for catalase in the matrix and for urate oxidase in the peroxisomal cores.
Other cell organelles should be totally negative.

3.5.6 Trouble Shooting

Many nonspecific gold particles over the cytoplasm:

-1) Check blocking and washing conditions.
-2) The antibody concentrations may be too high.
-3) Reduce the protein A-gold concentration or incubation time.

No gold particles in POs:

-1) Increase the antibody concentration.
-2) Use 0.05% to 2% Tween 20 in the incubation medium.
-3) Use protein A-gold with smaller gold particle size.

3.6 Embedding of Peroxisomal Fractions for Electron Microscopy

3.6.1 Aim

Visualization of the peroxidatic activity of catalase using the alkaline 3,3'-diaminobenzidine (DAB)-method (Fahimi 1968, 1969). Embedding of peroxisomal fractions with optimal morphological preservation.

3.6.2 Equipment

3.6.2.1 Access to

- an electron microscope (Hitachi H 600) for the demonstration of embedded PO-fractions
- an ultramicrotome (Reichert Ultracut S) and a knifemaker (Reichert Knifemaker II) with glass stripes
- a trimmer (Reichert Ultratrim)
- a precision balance
- a 4°C refrigerator
- an oven (60°C)

3.6.2.2 On the bench

Basic Equipment (see 3.3.2.2)

Additional Equipment
- a filtration apparatus (according to Baudhuin et al 1967)
- a gas tank with dry nitrogen and reduction valves
- a water bath shaker (37°C)
- a carrousel for incubating glass vials containing filterpieces
- a knife for dissection or razor blades
- a big glass (round ca. 30cm, height 10cm) with a glass cover (for overnight air-thight incubation of glass vials with Epon)
- 4-5ml glass incubation vials with caps
- plastic cups (ca. 50ml) for preparing Epon 812
- 1ml tuberculin-syringes
- Millipore filters (round, 13mm diameter and pore size 0.05µm)
- BEEM capsules (size 3) and capsule holders
- white paper, a pencil and a scissors
- white adhesive tape

3.6.3 Chemicals and Solutions

- density gradient medium with PO-banding density (for the exact protocol see section 1.1)
- 25% aqueous glutaraldehyde (GA) stock solution (ultrapure, Serva): dilute to 2.5%- and 1.25%-solutions in gradient medium; which gradient medium you will use depends on the PO-isolation procedure for the fixation of the PO-fractions.
- 2% osmium tetroxide solution (aqueous); always prepare the day before, since osmium dissolves very slowly.
- $K_4[Fe(CN)_6]$, 3 H_2O: 300 mg in 10 ml ultrapure water
- 0.2 M NaH-maleate buffer stock solution (maleic acid and NaOH), pH 5.2
- 2% uranyl acetate in 0.05M NaH-maleate buffer, pH 5.2
- Dimethylarsinic acid-Na-salt for preparing 0.1M Na-cacodylate buffer pH 7.4
- 0.05M Teorell-Stenhagen (TS) buffer containing 0.05M H_3PO_4, 0.057M boric acid, 0.035M citric acid and 0.345M NaOH (use for 100 ml: 352.4mg boric acid, 704mg citric acid (monohydrate), 1ml 5M H_3PO_4 and 345mM NaOH)
- 3,3'-diaminobenzidine (DAB; Sigma)
- 30% H_2O_2 (Merck)
- alkaline DAB-medium: 0.2% DAB, 0.15% H_2O_2, 0.01M Teorell-Stenhagen-buffer, pH 10.5
- ultrapure water
- absolute ethanol
- 1,2-ethoxypropane = propylene oxide
- agar agar, 3g/40ml H_2O; stir on a heating plate until the solution boils and subsequently cool down to ca. 37°C, with further stirring.
- all components for Epon 812 (MNA, DDSA, Epon 812, DMP-30); for 25ml Epon use: 8.1g, 6.1g, 13g and 0.375ml, respectively.
- $CaCl_2$ (crushed)
- acetone

3.6.4 Experimental Procedure

- For the isolation of highly purified PO-fractions see Experiment 1.1.
- Prepare the filtration apparatus before starting the fixation of the fractions; connect the tank with dry nitrogen to the system and insert the Millipore filters into the filtration device.
- Prepare two different fixing solutions, the first one containing 2.5% GA, the second one with 1.25% GA in appropriate gradient medium.
- You should always fix the POs in the same medium as that used for the gradients with the appropriate density, where they are banding.
- The volume that is to be used of the PO-fraction depends on their protein concentration; use 400µg protein (Lowry) for one Millipore-filter.
- Pipette the POs (400µg protein = X µl) into an Eppendorf tube.

51

- For fixation of the POs, add the same volume (X µl) of the 2.5% -GA- gradient medium to the PO-fraction (final concentration GA in the PO-solution: 1.25%); never use concentrated GA (25%) for adding to the POs; otherwise the organelles will clump together.
- Fill up the solution to 1 ml with 1.25% -GA-gradient medium.
- Load the whole solution into the filter holder, close the top and start the filtration by applying 1 atm pressure with N_2-gas.
- Filter the POs very slowly with low pressure (don't use more than 1 atm filtration pressure).
- The filtration of the whole solution takes about 15-20 min; do not exceed 30 min fixation time!
- In the meantime, prepare the alkaline DAB-incubation medium (always prepare freshly; DAB must be dissolved in water before adding the TS-buffer; the color of the solution should be rose-brown; stored it in the dark until use).
- Heat the water bath shaker to 37°C.
- After filtration, switch off the filtration pressure, open the filter holders, take out the filters with the fine forceps and fix the whole filters (PO-side up) with 1.25% -GA-gradient medium in glass vials for another 10-15 min; don't exceed 45 min fixation time.
- Remove the fixative and wash the filters 3 times with 0.1M Na-cacodylate buffer, pH 7.4; be sure not to pipette directly onto the fractions.
- Prepare a second glass vial with a different label on it.
- Take out the filters with the forceps (only touch the rim of the filter) and cut each into two halfs with a razor blade or a dissection knife; don't let the filters dry!
- Proceed with the two filter halves as different preparations: you will use one for routine electron microscopy; the other half will be processed for the cytochemical localization of catalase.
- Prepare the agar solution (1 g/40 ml H_2O), so that it is ready to use after the postfixation of the samples with osmium tetroxide.

Proceed as follows: use two timers!

1	2
Leave one half of the filter with isolated POs in Na-cacodylate buffer on ice (will be processed for routine electron microscopy).	Wash the other half three times with 0.01 M TS-buffer (will be processed for DAB cytochemistry).
|	|
|	|
|	|
|	|

	3
	Incubate the filter half with the DAB-medium for 1 h at 37°C in darkness; use gentle shaking of the water bath shaker.
4	
Remove the Na-cacodylate buffer and add 2% aqueous osmium tetroxide (only cover the surface of the filter; always use a ventilated hood and gloves for handling osmium!)	
5	
After 30 min add the same volume of $K_4[Fe(CN)_6)]$ solution, and reduce osmium for another 30 min.	6
	Wash three times with Na-cacodylate buffer.
	7
8	Add 1% aqueous osmium tetroxide; postfix for 1h at 4°C.
Remove the reduced osmium (collect and dispose of as toxic waste), wash with 0.05M NaH-maleate buffer, pH 5.2 and add the 2% uranyl acetate solution; leave for 30 min at 4°C.	
9	10
Wash 3 times with NaH-maleate buffer; take out the filters and cut them into long stripes (they should fit into the BEEM capsules later!).	Remove the osmium, wash three times with NaH-maleate buffer and add 2% uranyl acetate solution; leave for 30 min at 4°C.
- Drop some agar onto a clean glass plate and insert the filter pieces into the agar; be sure that the filters will not dry and do not enclose air bubbles!	11
	Proceed as indicated in 9.

- Cut the agar as indicated:

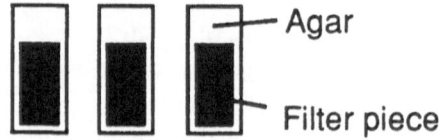

- Cut the surrounding agar as small as possible.
- It is necessary to enclose the filter pieces in agar, since the synthetic filters will dissolve in propylene oxide.
- Store the pieces in 75% ethanol on ice until the pieces with DAB are ready; when the filter pieces of both groups are in 75% ethanol, process the different groups at the same time.
- Dehydrate both preparations in series of graded ethanol: 3 times, 15 min each 75% -, 85% -, 95% - and absolute ethanol.
- In the meantime, prepare the Epon 812 mixture: add first MNA, DDSA and Epon 812, and stir for 30 min; then add DMP-30 (the mixture will change its color to orange!) and stir for another 30 min before use.
- Proceed with the samples for embedding: change 3 times 15 min propyleneoxide.
- Two times 30 min propyleneoxide/Epon-mixture (v/v)
- Add pure Epon-mixture and leave the samples on a caroussel overnight at room temperature to mix them well.

The Next Morning:

- Embed the filterpieces in BEEM capsules with Epon-mixture.
- Always use freshly prepared Epon for the embedding! For 20 small BEEM-capsules, you need ca. 5 ml Epon.
- While stirring of the Epon-mixture, prepare the BEEM-capsules; insert small numbers on paper in the tops of each capsule with a pair of forceps as indicators for the different groups.
- Put the capsules into the holder.
- Fill the capsules with 3 drops of Epon-mixture; use a 1ml tuberculine-syringe to add the Epon.
- Take out the filter-agar-pieces from the glass vials with forceps; remove the old Epon with filter paper before the insertion of the pieces into the BEEM capsules.
- Insert the agar-filter-pieces into the capsules, with the filter in the direction of the bottom of the capsule.
- Fill up the capsules with Epon-mixture and be sure that no air bubbles will stick to the filters; remove them with forceps. Otherwise, you will make holes in the block.

- Polymerize the capsules with Epon 812 for 2 days at 60°C in a polymerization oven; place a Petri dish with $CaCl_2$ into the oven to dessicate the air in the oven.

After polymerization, the blocks are ready to be cut. POs are easily found in semithin sections, if the sections are stained with Richardsons stain. Trim the blocks containing the filters in a diagonal position for ultrathin-sectioning.

3.6.5 Expected Results

In semithin sections (after Richardsons stain), the peroxisomal fraction will be visible even in the normal light microscope as a dark blue stripe. The agar will be stained only very faintly blue, and the area of the filters will be colorless.

Individual POs are not distinguishable!

After ultrathin-sectioning with a diamond knife, the fractions may be examined by electron microscopy.

Ten to twenty layers of POs will be observed on the filters.

The POs should be well preserved, with a fine granular matrix and the cores in the middle of the organelle. In DAB-incubated preparations, POs will be stained intensely. DAB forms a very electron dense, homogeneous precipitate in the matrix of the POs. The cores will appear as negative areas inside the matrix. Only rarely extracted particles or free cores should be seen.

3.6.6 Trouble Shooting

Several artifacts may arise during the embedding of the fractions for electron microscopy. If a peroxisomal fraction is not well preserved (as seen in electron microscopy) it doesn't automatically mean that the biochemical isolation procedure wasn't appropriate. Always keep in mind that POs are very fragile organelles and may be altered during fixation and embedding for electron microscopy.

If POs are compacted together, the filtration pressure must have been too high during filtration onto the Millipore filters, or the filter pores were clogged; never load too much protein onto the filters.

If POs are extracted, the density of the fixative-solution might have been inappropriate; it should have the density where POs band. Never use buffers for fixation without gradient material.

Never use old or frozen preparations of POs for embedding, since they will be terribly extracted.

Pelleting instead of filtration of the POs will lead to many artifacts; pelleting of POs will never yield a good PO-morphology!

References

1. Angermüller S, Fahimi HD (1986) Ultrastructural cytochemical localization of uricase in peroxisomes of rat liver. J Histochem Cytochem 34:159-165
2. Baudhuin P, Evrard P, Berthet J (1967) Electron microscopic examination of subcellular fractions. I. The preparation of representative samples from suspensions of particles. J Cell Biol 32:181-191
3. Baumgart E, Stegmeier K, Schmidt FH, Fahimi HD (1987) Proliferation of peroxisomes in pericentral hepatocytes of rat liver after administration of a new hypocholesterolemic agent (BM 15766). Sex-dependent ultrastructural differences. Lab Invest 56:554-564
4. Baumgart E, Völkl A, Hashimoto T, Fahimi HD (1989) Biogenesis of peroxisomes: Immunocytochemical investigation of peroxisomal membrane proteins in proliferating rat liver peroxisomes and in catalase-negative membrane loops. J Cell Biol 108:2221-2231
5. Baumgart E, Völkl A, Fahimi HD (1990) Proliferation of peroxisomes without simultaneous induction of the peroxisomal fatty acid ß-oxidation. FEBS Lett 264: 5-9
6. Beier K (1992) Light microscopic morphometric analysis of peroxisomes by automatic image analysis: advantages of immunostaining over the alkaline DAB method. J Histochem Cytochem 40:115-121
7. Beier K, Fahimi HD (1992) Application of automatic image analysis for quantitative morphological studies of peroxisomes in rat liver in conjunction with cytochemical staining with 3,3'-diaminobenzidine and immunochemistry. Micr Res Tech 21:271-282
8. Beier K, Völkl A, Hashimoto T, Fahimi HD (1988) Selective induction of peroxisomal enzymes by the hypolipidemic drug bezafibrate. Detection of modulations by automatic image analysis in conjunction with immunoelectron microscopy and immunoblotting. Eur J Cell Biol 46:383-393
9. De Duve C, Baudhuin P (1966) Peroxisomes (microbodies and related particles). Physiol Rev 46:323-357
10. Fahimi HD (1968) Cytochemical localization of peroxidase activity in the rat hepatic microbodies (peroxisomes). J Histochem Cytochem 16:547-550
11. Fahimi HD (1969) Cytochemical localization of catalase in rat hepatic microbodies (peroxisomes). J Cell Biol 43:275-288
12. Fahimi HD, Baumgart E (1992 in press) Peroxisomes. In: Ogawa K and Barka T (eds) Electron Microscopic Cytochemistry and Immunocytochemistry in Biomedicine. Boca Raton: CRC Press
13. Fahimi HD, Reinicke A, Sujatta M, Yokota S, Oezel M, Hartig F, Stegmeier K (1982) The short- and long-term effects of bezafibrate in the rats. Ann NY Acad Sci 386:111-135
14. Fahimi HD, Sies H (1987) Peroxisomes in Biology and Medicine. Heidelberg: Springer-Verlag, pp. 1-458.
15. Fringes B, Reith A (1982) Time course of peroxisome biogenesis during adaptation to mild hyperthyroidism in rat liver. A morphometric/stereologic study by electron microscopy. Lab Invest 47:19-26
16. Gorgas K, Krisans S (1989) Zonal heterogeneity of peroxisome proliferation and morphology in rat liver after gemfibrozil treatment. J Lipid Res 30:1859-1875

17. Hartl F-U, Just WW (1987) Integral membrane polypeptides of rat liver peroxisomes: topology and response to different metabolic states. Arch Biochem Biophys 255:109-119
18. Hruban Z, Vigil EL, Slesers A, Hopkins E (1972) Microbodies. Constituent organelles of animal cells. Lab Invest 27:184-191
19. Litwin JA, Völkl A, Stachura J, Fahimi HD (1988) Detection of peroxisomes in human liver and kidney fixed with formalin and embedded in paraffin: the use of catalase and lipid ß-oxidation enzymes as immunocytochemical markers. Histochem J 20:165-173
20. Litwin JA, Yokota S, Hashimoto T, Fahimi HD (1984) Light microscopic immunocytochemical demonstration of peroxisomal enzymes in epon sections. Histochemistry 81:15-22
21. Lüers G, Beier K, Hashimoto T, Fahimi HD, Völkl A (1990) Biogenesis of peroxisomes: sequential biosynthesis of the membrane and matrix proteins in the course of hepatic regeneration. Eur J Cell Biol 52:175-184
22. Osmundsen H, Bremer J, Pedersen JI (1991) Metabolic aspects of peroxisomal ß-oxidation. Biochim Biophys Acta 1085:141-158
23. Reddy JK, Lalwani ND (1983) Carcinogenesis by hepatic peroxisome proliferators: evaluation of the risk of hypolipidemic drugs and industrial plasticizers to humans. CRC Crit Rev Toxicol 12:1-58
24. Slot JW, Geuze HJ (1985) A new method for preparing gold probes for multiple-labeling cytochemistry. Eur J Cell Biol 38:87-93
25. Völkl A, Baumgart E, Fahimi HD (1988) Localization of urate oxidase in the crystalline cores of rat liver peroxisomes by immunocytochemistry and immunoblotting. J Histochem Cytochem 36:329-336
26. Völkl A, Fahimi HD (1985) Isolation and characterization of peroxisomes from the liver of normal untreated rats. Eur J Biochem 149:257-265
27. Yamamoto K, Fahimi HD (1987) Biogenesis of peroxisomes in regenerating rat liver. I. Sequential changes of catalase and urate oxidase detected by ultrastructural cytochemistry. Eur J Cell Biol 43:293-300
28. Yamamoto K, Fahimi HD (1987) Three-dimensional reconstruction of a peroxisomal reticulum in regenerating rat liver: evidence of interconnections between heterogeneous segments. J Cell Biol 105:713-722
29. Zaar K, Fahimi HD (1991) Immunoelectron microscopic localization of the isozymes of L-α-hydroxyacid oxidase in renal peroxisomes of beef and sheep: evidence of distinct intraorganellar subcompartimentation. J Histochem Cytochem 39:801-808
30. Zaar K, Völkl A, Fahimi HD (1991) Purification of marginal plates from bovine renal peroxisomes: identification with L-α-hydroxyacid oxidase B. J Cell Biol 113:113-121

Acknowledgement

The organizers wish to thank Dr J.P. ZHAND for his help in electron microscopy during the FEBS practical course.

II Molecular Biology

4 Northern Blotting Analysis of Rat Liver mRNA Encoding for Peroxisomal Acyl-CoA Oxidase

M. CHERKAOUI MALKI and F. CAIRA

4.1 Introduction and Aims

Several compounds, which encompass a number of fibrate molecules and analogues, have been classified as peroxisome proliferators. Many of these fibrates, with hypolipemic action, are employed as human drugs in the prevention of arteriosclerosis [2]. In rodents, peroxisome proliferators produce a pleiotropic response, including hepatomegaly and an increase in the number of peroxisomes and in the activity of several enzymes. These enzymes are primarily located in the peroxisomes, particularly those of the long chain fatty acid β-oxidation system [2].

Acyl-CoA oxidase is the rate-limiting enzyme of the peroxisomal β-oxidation system [5]. A large increase in this enzyme's transcription rate is observed in liver from rats treated by peroxisome proliferators [3, 6].

The aim of this experiment is to detect by the Northern blotting method, at the post-transcriptional level, changes in the acyl-CoA oxidase mRNA rate in rat liver after exposure to a peroxisome proliferator.

Three steps are required to accomplish Northern blotting. First, total RNA, isolated from the desired tissue or cells [1], is size-separated by agarose gel electrophoresis under denaturing conditions [7]. Then separated RNA is transferred to a more suitable solid support (nitrocellulose or nylon membranes). Finally, the specific mRNA is detected using a ^{32}P-labelled complementary sequence [4]. Such a probe is a fragment of cloned DNA or a synthetic oligonucleotide. The DNA is hybridized to the specific mRNA under conditions that favor formation of an RNA-DNA duplex. The intensity of the band on the X-ray film is proportional to the homologous RNA present in the studied tissue.

4.2 RNA Preparation

Great care should be taken for RNA preparation and handling because analysis of RNA requires the absence of RNAase contamination in the purified RNA.

General laboratory glassware should be baked at 180°C overnight. Sterile disposable plasticware is used in all experiments. Except when it is inadvisable, centrifuge tubes are sterilized by autoclaving, and the electrophoresis apparatus is rinsed in sterile water, dried with ethanol and immersed in 3% H_2O_2. After 10 min, the apparatus is rinsed extensively with sterile water. Investigators should wear disposable gloves during manipulation, including preparation of materials and solutions.

4.2.1 Equipment, Biological Material and Solutions

4.2.1.1 Equipment

Access to
- a low speed centrifuge (Beckman GPKR) and a rotor for 50 ml tubes (Beckman GH-3.7)
- a high speed centrifuge (Beckman J2-21), a 8 x 50 ml rotor (Beckman JA-20) and 50 ml sterile, capped polyallomer tubes (Beckman 357001)
- a motor-driven homogenization device for tissue grinders
- a refrigerated microcentrifuge for Eppendorf tubes (Sigma 202-MK)
- a rotary evaporator type Speed-Vac (Jouan RC 1010)
- a tissue homogenizer (Ultraturrax S 25 - G10)
- a spectrophotometer (Kontron series Uvikon)
- a balance
- 4°C and -70°C refrigerators
- a fume hood
- a guillotine

On the Bench
- aluminum foil
- adjustable automatic pipettes and sterile tips
- a dissection board (forceps and scissors)
- Parafilm
- pipette pumps
- disposable respirators with exhalation valve and an adjustable nosepiece (Sigma)
- serological pipettes (10 ml)
- 1 ml quartz microcuvettes for spectrophotometry
- sterile, disposable centrifuge tubes (50 ml, Nunc) with racks
- sterile Eppendorf tubes and racks
- sterile glass Pasteur pipettes
- sterile latex surgical gloves
- sterile polyethylene squeeze pipettes (Movettes)
- a 30 ml glass Potter-Elvehjem tissue grinder with a fitted Teflon pestle
- a vortex mixer

4.2.1.2 Biological Material

Rats were treated with ciprofibrate (3 mg/kg body weight) for 1 week. Control and treated rats were starved for one day before experiments.

4.2.1.3 Solutions
- chloroform : isoamyl alcohol mixture (24:1, v/v)
- 80% ethanol
- ethanol 95°
- 3M LiCl, 6M Urea
- 3M Na acetate, pH 6.0
- 0.1 N NaOH
- 10% SDS
- TE buffer : 10 mM Tris-HCl pH 8.0, 1 mM EDTA
- TES buffer : 10 mM Tris-HCl pH 8.0, 1 mM EDTA and 0.5 % SDS
- salt-saturated phenol (SS-phenol) : phenol saturated with TE
- SS-phenol : chloroform : isoamyl alcohol mixture (25:24:1, v/v)
- Ultrapure and sterile water (MilliQ, Millipore)

4.2.2 Experimental Procedure

Day 1 (1 h)

[1] Decapitate the rats using the guillotine. Wipe abdomen skin with ethanol and cut open the abdomen. Excise the organ of interest (*i.e.* liver) and cut into large pieces on alimunium foil. Weigh 2 g and transfer the tissue to the 50 ml centrifuge tube.

[2] Add 10 ml of ice-cold 3 M LiCl, 6 M urea solution. Homogenize each piece of tissue with the Ultra-Turrax homogenizer on ice at 20,000 rpm for 4 cycles of 15 sec.

[3] Transfer the mixture to the 30 ml Potter-Elvehjem grinder. Homogenize by 6 strokes at 1500 rpm, and then transfer the foaming homogenate to the polyallomer centrifuge tubes.

[4] Complete to 10 ml/g of tissue in each tube with ice-cold 3 M LiCl, 6 M urea solution and store on ice for 5 min.

[5] Add 10 % SDS to reach 0.1 % SDS in the homogenate. Mix by shaking and leave overnight at 4°C.

Day 2 (Morning)

[6] Pellet the RNA by centrifugation in the high speed centrifuge at 10,000 rpm (12,000 g) for 30 min.

[7] Remove the supernatant and add TES buffer, 10 ml/g of tissue. Suspend the pellet completly by vortexing.

[8] Transfer to a 50 ml Nunc tube and extract with an equal volume of SS-phenol. Separate the two phases by centrifugation for 5 min at 3000 rpm (2000 g) in the low speed centrifuge.

[9] Transfer the upper aqueous phase to a new Nunc tube. Add one volume of SS-phenol. Mix by shaking. Spin for 5 min at 3000 rpm.

[10] Transfer the upper aqueous phase to a new Nunc tube. Add one volume of SS-phenol:chloroform:isoamyl alcohol mixture. Mix by shaking and spin for 5 min at 3000 rpm.

[11] Repeat twice that indicated in paragraph [10].

[12] To the aqueous phase, add 1 volume of chloroform:isoamyl alcohol. Mix by shaking and spin for 5 min at 3000 rpm.

[13] Transfer the upper aqueous phase to a new Nunc tube. Add 1/10 volume of 3 M sodium acetate, pH 6.0. Mix and complete with 2.2 volumes of cold absolute ethanol. Vortex and store for 20 min at - 70 °C.

[14] Spin for 10 min at 3000 rpm.

[15] Resuspend the pellet in 15 ml of 80 % ethanol and spin for 10 min at 3000 rpm. Wash the pellet two additional times with 80 % ethanol.

[16] Evaporate the rest of ethanol off the pellet in the vaccum rotary apparatus, but do not dry completly.

[17] Resuspend each pellet in 200 to 400 μl of sterile H_2O.

[18] Determine RNA quantities by measuring optical density (OD) at 260 nm. For RNA, one unit of OD corresponds to 40 μg/ml.

4.3 Analysis of RNA by Northern Blotting

4.3.1 Equipment, Chemicals and Solutions

4.3.1.1 Equipment

Access to
- a hybridization incubator with 20 ml hybridazation tubes
- a Polaroid MP-4 system with a yellow-UV filter and a UV transilluminator (305 nm), in a dark room
- a microwave oven
- a refrigerator at 4°C
- a refrigerated microcentrifuge (Sigma 202-MK)
- a thermostated incubator for Eppendorf tubes
- a thermostated water bath with rack for Eppendorf tubes
- a bag sealer
- a fume hood

On the Bench
- adjustable automatic pipettes and sterile tips
- a disposable scalpel
- a horizontal electrophoresis unit and accessories (Hoefer HE 33)
- a power supply (500 volts)
- a gel scoop
- heat-seal bags
- forceps for handling filters (Millipore)
- Polaroid 667 films
- Nylon membranes for RNA transfer
- Parafilm
- pipette pumps
- disposable respirators with an exhaltation valve and an adjustable nosepiece
- a 30 cm length ruler
- a safelight lamp
- safety glasses for UV
- Saran wrap
- serological pipettes (10 ml)
- sterile Eppendorf tubes and racks
- sterile glass cylinders : 50 ml, 250 ml and 500 ml
- sterile latex surgical gloves
- a timer
- a tray (20 x 50 cm)
- two glass plates (20 x 30 cm) and a weight (500g)
- a vortex
- Whatman 3 MM filter paper

4.3.1.2 Chemicals and Solutions

Chemicals
- agarose, molecular biology quality (Sigma)
- formaldehyde (Fluka)
- RNA molecular weight markers for electrophoresis (Boehringer)

Solutions
- 0.1 mg/ml ethidium bromide
- deionized formamide
- 40 % glycerol, 0.25% bromophenol blue
- 10 x MOPS buffer : 0.2 M MOPS (3-(N-morpholino)propansulfonic acid), pH 7.0, 80 mM Na acetate, 10 mM EDTA
- hybridization buffer (Rapid-Hyb, Amersham)
- 10 mg/ml salmon sperm DNA
- 20 x saline sodium citrate (SSC) buffer: 3 M NaCl, 0.3 M Na$_3$ citrate, 2 H$_2$O, pH 7.0 in sterile water.
- 10 x SSC and 2 x SSC
- 2 x SSC, 0.1% SDS
- 0.1 x SSC, 0.1% SDS
- sterile ultrapure water (MilliQ, Millipore)

4.3.2 Experimental Procedure

Day 2 (Evening)

[1] Using the microwave oven, boil 0.6 g agarose in 43 ml water for a 6 x 8 cm, 1 % gel. Let cool to 50°C and add, in a fume hood, 6 ml 10 x MOPS buffer and 10.8 ml formaldehyde. Mix by swirling.

[2] In a fume hood, cast the gel, and allow it to set for at least 20 min at room temperature. The gel is submerged in 200 ml running buffer containing 20 ml 10 x MOPS, 36 ml formaldehyde and 144 ml ultrapure water.

[3] Prepare the samples by mixing the following in a sterile Eppendorf tube:

- 10 x MOPS	1.0 µl
- formaldehyde	3.5 µl
- formamide	10.0 µl
- 40% glycerol, 0.25 % bromophenol blue	2.0 µl
- ethidium bromide	3.0 µl
- RNA (1- 25 µg)	5.0 µl

Heat at 65°C for 15 min to denature the RNA, and then chill the samples in ice. Spin for 5 sec in a microcentrifuge at 4°C.

[4] Load the samples into the wells of the gel. Run the gel at 5 V/ cm for ~2 h, until the bromophenol blue dye migrates up to three-fourths of the gel.

[5] Rinse the gel several times with water, and drain. Photograph the gel (with a ruler) on the UV transilluminator using the Polaroid MP-4 system.

[6] Place the gel in 200 ml of 10 x SSC for 10 min with continuous shaking. Rinse the gel an additional time with the same solution.

[7] For the capillary transfer assembly, see Figure 1. Fill the tray with 400 ml of 10 x SSC and wet a piece of Whatman 3 MM filter paper (20 x 40 cm) in 10 x SSC buffer.

[8] Over the tray, place a glass plate (20 x 30 cm) and drape the wetted Whatman filter paper over the plate. The ends of the filter paper, hanging into the buffer, act as blotting paper. Smooth out all air bubbles with a sterile plastic pipette.

[9] Using gloves and Millipore forceps, cut a piece of Nylon filter (6.2 x 8.2 cm) and wet the filter on the surface of a dish of sterile water. Then, immerse the filter in 10 x SSC for 5 min. Cut two pieces of Whatman 3 MM filter paper (6 x 8 cm) and wet them in 2 x SSC.

[10] Place the gel on the glass plate, the sample wells opening downward, and smooth out any air bubbles. Lay the Nylon filter on the gel and smooth. Place two pieces of Whatman 3 MM filter paper on top of the Nylon filter and smooth again. Put on the stack of dry paper towels. Cover with a glass plate and place a small weight (500 g) on the transfer pyramid.

[11] Capillary transfer takes about 12 - 16 h at room temperature.

Day 3 (Morning)

[12] Take off the stack of paper towels and Whatman 3 MM filter paper. Mark the Nylon filter with a marker in well areas to indicate the electrophoretic origin. Cut one corner of the Nylon filter and gel to mark filter orientation.

[13] Remove the Nylon filter(= blot), wash 5 min in 2 x SSC and dry the blot with a hair drier. Crosslink RNA to the Nylon filter (in Saran wrap) by exposure, RNA side down, to UV light for 3 min at 305 nm.

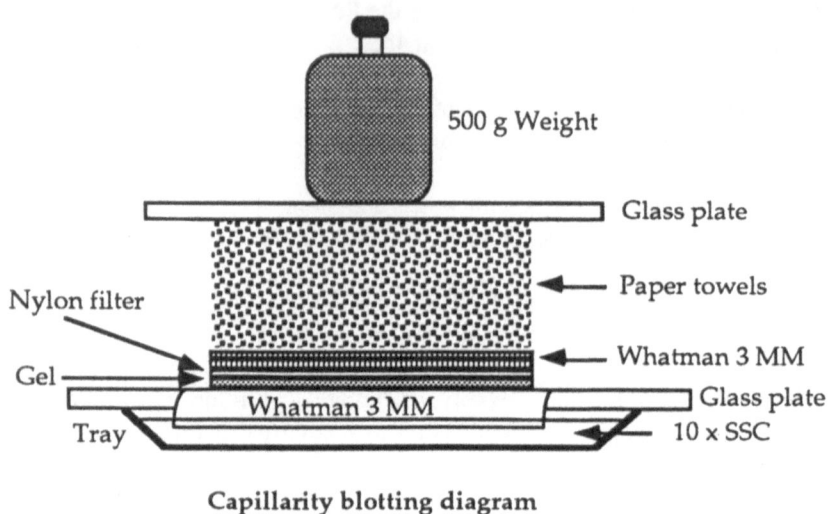

Capillarity blotting diagram

Figure 1: Northern blotting appparatus set-up.

[14] Transfer the blot to a hybridization tube, and add 4 ml of Rapid-Hyb buffer. Seal with a cap and let prehybridize for 60 min at 65°C.

[15] Boil the radioactive probe for 10 min in a screw-cap tube. Transfer to ice .

[16] Add the probe solution to the hybridization tube. Incubate for 2.5 h at 65°C in the hybridization incubator.

[17] Rinse the filter twice at room temperature with 50 ml of 2 x SSC, 0.1% SDS, and wash with the same solution (100 ml) at room temperature for 20 min, followed by two washes at high-stringency with 100 ml of 0.1 x SSC, 0.1% SDS (prewarmed to 65°C), 25 min for each wash at 65°C.

[18] Seal the filter in a heat-seal bag and expose it overnight to a X-ray film (see 4.6).

4.4 Rapid Multiprime DNA Labeling

Using a random sequence of hexanucleotides as a primer and the Klenow fragment of DNA polymerase I, denatured DNA can be radiolabelled from various sources with [α-^{32}P] deoxyribonucleoside-5'-triphosphate. This protocol is based on the method of Feinberg and Vogelstein [4], as described by the commercial supplier.

4.4.1 Equipment, Chemicals and Solutions

4.4.1.1 Equipment

Access to
- an automatic scintillation counter (Beckman LS 6000IC)
- a freezer (-20 °C)
- a microcentrifuge for Eppendorf tubes
- a thermostated water bath with a rack for Eppendorf tubes

On the Bench
- adjustable automatic pipettes plus sterile tips
- biohazard disposal bags and a container
- a 10 ml automatic liquid dispenser
- scintillation counting vials
- a Geiger counter for β-emitters
- Parafilm
- sterile Eppendorf tubes and racks
- sterile latex surgical gloves
- an acrylic storage rack for Eppendorf tubes containing a β-emitter
- a timer and a vortex mixer
- an acrylic workshield for β-emitters and spill trays

4.4.1.2 Chemicals and Solutions

- Prime-a-Gene® system kit (Promega)
- [α-^{32}P] dCTP (3000 Ci/mmol, Amersham, 250 μCi)
- 0.5 M EDTA pH 8.0
- Scintillation solution (Pico-Fluor^{TM} 15, Packard)
- Sterile ultrapure water (MilliQ, Millipore)

4.4.2 Experimental Procedure

Day 2 (Evening)

[1] Heat the DNA to be labeled to 95°C for 10 min in a boiling water bath and chill on ice.

[2] Prepare the following reaction mixture in an Eppendorf tube on ice:

- 29 µl sterile water
- 5 µl DNA (25 ng)
- 2 µl dNTP mixture minus dCTP
-10 µl random hexamers in labelling buffer
- 5 µl 3000 Ci/mmol [α-^{32}P] dCTP (50 µCi)
- 1 µl Klenow fragment (5 units)

Mix by pipetting up and down, spin in the microcentrifuge and incubate at 37°C for 30 min.

[3] Stop the reaction with 2 µl 0.5 M EDTA pH 8.0, or by heating to 95°C for 5 min.

[4] Remove unincorporated radioactive nucleotides ([α-^{32}P] dCTP) from the labeled DNA by chromatography on a Sephadex G-50 column (see 4.5).

[5] Determine the specific activity of the DNA probe on a 1 µl aliquot (2-5 10^8 cpm/µg DNA).

4.5 Gel-filtration Chromatography on Sephadex G-50

4.5.1 Equipment, Chemicals and Solutions

4.5.1.1 Equipment

Access to
- a low speed centrifuge (Beckman GPKR) and a swinging bucket rotor (Beckman GH-3.7)

On the Bench
- a 10 ml automatic liquid dispenser
- 15 ml plastic tubes for centrifugation (Falcon n° 2095)
- a Geiger counter for β-emitters
- forceps
- Parafilm
- pipette pumps and serological pipettes (10 ml)
- sterile disposable syringes (2 ml)
- sterile latex surgical gloves
- sterile siliconised glass wool
- a timer

4.5.1.2 Chemicals and Solutions

- Labelled DNA probe in its labelling medium (54 µl) (see 4.4.2)
- TE buffer : 10 mM Tris-HCl pH 8.0, 1 mM EDTA
- Sephadex G-50 Medium (Sigma)
- Scintillation solution (Pico-Fluor^{TM} 15, Packard)

4.5.2 Experimental Procedure

4.5.2.1 Preparation of Sephadex G-50

Sephadex G-50 is prepared as indicated by Maniatis et al. (1989): in a 250 ml bottle, wash 5 g of Sephadex G-50 several times with 100 ml sterile water and equilibrate the resin in TE buffer pH 8.0; autoclave at 0.7 atm (114°C) for 15 min.

4.5.2.2 Gel Filtration Chromatography

[1] Plug the bottom of a 2 ml sterile disposable syringe, using the syringe pestle, with sterile siliconized glass wool. Rinse the syringe with 1 ml TE buffer and pack the glass wool with the pestle. Add Sephadex G-50 prepared in TE buffer until the syringe overflows and let the resin shrink for 2 min. Fill the column again with resin until it overflows and take off the excess of resine with a spatula.

[2] Insert the column into a 15 ml Falcon tube. Spin for 5 min at 1600 g, at room temperature, in the GPKR centrifuge. Add resine in suspension, and recentrifuge.

[3] Insert the column into another 15 ml Falcon tube; add 130 µl TE to the mixture (54 µl) containing labelled DNA probe (see 4.4.2). Mix and carefully apply the DNA sample. Spin as in step [2].

[4] Discard the syringe in a radioactive waste disposal bag. Transfer the effluent (~150 µl) to a sterile Eppendorf tube and measure the volume. Store at -20°C until needed.

[5] Add 1 µl of purified labelled DNA to 10 ml Pico-Fluor in a counting vial. Count the radioactivity incorporated into the DNA probe.

4.6 Autoradiography

4.6.1 Equipment and Chemicals

4.6.1.1 Equipment

Access to
- a darkroom

On the Bench
- a developing tray
- a 20 x 25 cm X-ray film (Kodak XAR-5)
- an exposure cassette with an intensifying screen
- a Geiger counter
- latex surgical gloves
- Polaroid 667 films and film hangers
- a safelight lamp
- Saran wrap
- an acrylic workshield for β-emitters and spill trays

4.6.1.2 Chemicals :

- processing chemicals : Dektol developer, GBX fixer and replenisher (Kodak)

Take care to maintain the temperature, which should be 18 - 22°C, of all solutions.

4.6.2 Experimental Procedure

[1] Bring the hybridized filter (see 4.3, [18]) in contact with the X-ray film in an exposure cassette with an intensifying screen. Expose overnight at -70°C

[2] Once the exposure time is over, quickly remove the X-ray film from the exposure cassette and immerse it immediately in GBX developer solution for about 1 min.

[3] Remove the film from the GBX developer and immerse it in GBX fixer solution for 5 min.

[4] Immerse the film in running water for at least 10 min.

[5] Dry the film at room temperature.

Acknowledgements

The authors wish to thank Professor Norbert Latruffe for his advice and encouragement, Dr. Takashi Osumi for the gift of peroxisomal acyl-CoA oxidase cDNA and Sterling-Winthrop Drug Laboratories for ciprofibrate supply.

References

1. Auffray C, Rougeon F (1980) Purification of mouse immunoglobulin heavy-chain messenger RNAs from total myeloma tumor RNA. Eur J Biochem 107:303-314
2. Cherkaoui Malki M, Bardot O, Lhuguenot JC, Latruffe N (1990) Expression of liver peroxisomal proteins as compared to other organelle marker enzymes in rats treated with hypolipidemic agents. Biol Cell 69:83-92
3. Cherkaoui Malki M, Lone YC, Corral-Debrinski M, Latruffe N (1990) Differential proto-oncogenes mRNA induction from rats treated with peroxisome proliferators. Biochem Biophys Res Commun 173:855-861
4. Feinberg AP, Vogelstein B (1983) A technique for radiolabelling DNA restriction endonuclease fragments to high specific activity. Anal Biochem 132:6-13
5. Lazarow PB, De Duve C (1976) A fatty acyl-CoA oxidizing system in rat liver peroxisomes; enhancement by clofibrate, a hypolipidemic drug. Proc Natl Acad Sci USA 73:2043-2046
6. Osumi T, Hijikata M, Ishii N, Myazawa S, Hashimoto T (1987) Cloning and structural analysis of the genes for peroxisomal ß-oxidation enzymes. In: Fahimi HD and Siess H (eds) "Peroxisomes in Biology and Medicine," Springer-Verlag, pp. 105-114
7. Sambrook J, Fritsch EF, Maniatis T (1989) in Molecular Cloning, Cold Spring Harbor Laboratory Press

5 Transient Co-transfection Assay: Measure of the Chloramphenicol Acetyl Transferase (CAT) Activity in Cytosol Extracts from Transfected Cell

O. BARDOT

5.1 Introduction and Aims

Transfection is a common procedure used to introduce DNA into mammalian cells. The use of transient transfection, in which the transfected cells are analysed only a few days after the introduction of DNA, allows the study of gene expression *in vivo*. The transcription of the transfected gene can be analysed by simply harvesting the transfected cells. This system uses a reporter plasmid to measure promoter activity. A reporter plasmid consists of the promoter of interest directing the synthesis of a protein that can be assayed for easily. The amount of this protein under different conditions reflects the ability of the promoter to initiate transcription.

The most commonly used protein is the enzyme chloramphenicol acetyl transferase (CAT). This bacterial enzyme catalyses the transfer of the acetyl group from acetyl-CoA to chloramphenicol, creating acetyl-chloramphenicol. The acetyl-chloramphenicol derivatives can be labelled using [^{14}C] chloramphenicol and then separated from free [^{14}C] chloramphenicol by thin-layer chromatography (TLC). Reporter plasmid vectors have been created and contain the CAT enzyme coding sequence, mammalian processing signals (polyadenylation signal) and polycloning sites where the promoter under study can be inserted [1, 2].

Recently identified by Issemann and Green [4], the peroxisome proliferator activated receptor (PPAR) recognises a response element (PPRE) in the upstream sequence of the rat acyl CoA oxidase gene (ACO). The PPAR is a nuclear hormone receptor and a ligand-activated transcription factor (fig. 1) [4]. In order to demonstrate the presence of a response element in the 5' flanking sequence of the ACO gene and examine the activation by a peroxisome proliferator, transient co-transfection assays have been performed. The 5' flanking sequence of the ACO gene has been cloned upstream from the rabbit β-globin promoter (G in Fig.1) and the CAT enzyme coding sequence. This reporter plasmid is called pACO (-1273/-471).G.CAT. A second plasmid, pACO (-1273/-582).G.CAT, has been constructed by deletion of PPRE, located 570 bp upstream from the start of transcription [5].

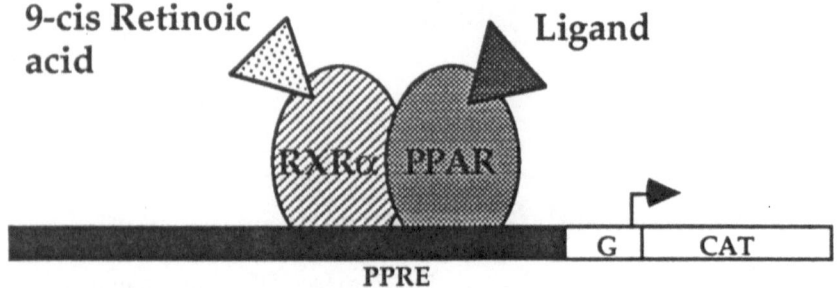

Figure 1: Schematic activation of CAT transcription by the heterodimer PPAR/RXRα activated by their ligands. RXRα : retinoïd-X-receptor α.

Two reporter plasmids have therefore been prepared:

- pACO (-1273/-471).G.CAT, which has the PPRE,
- pACO (-1273/-582).G.CAT, which does not have the PPRE.

These reporter plasmids have been co-transfected into the mouse Hepa 1 cell line (derived from a hepatoma) with the PPAR expression vector, pSG5-PPAR, that encodes the receptor [3]. The co-transfection of the different plasmids is shown in Table 1. In this co-transfection, 10 μg of total plasmid DNA were transfected. pBSM (Bluescript from Stratagene) was used as a carrier, the expression vector pSG5 [3] was used as the negative control for the expression of the receptor, and pCH110 was a reporter plasmid containing the β-galactosidase gene (Pharmacia). This last plasmid was used as an internal control to determine the transfection efficiency. To activate PPAR, the cells were treated with a potent peroxisome proliferator, Wyeth 14,643 (10^{-5}M), for 48 hours and then harvested. Cytosol extracts were prepared by freeze-thaw technique, and β-galactosidase activity were assayed [1]. In order to correct the variation in transfection efficiency, the same number of β-galactosidase units was used for the CAT assay. In this way, the percentage of CAT acetylation is normalized to reflect the level of transcription from the reporter plasmid. This percentage can be used to deduce the existence and the activity of the PPRE in the sequence under study. In the present protocol, we will describe the measurement of the CAT activity in cytosol extracts from transfected cells.

Plates #	pBSM	pSG5	pSG5-PPAR	pCH 110	pACO(-1273/-471).G.CAT	pACO(-1273/-582).G.CAT	Wyeth -14,643
1/2	5[a]	1	-	3	1	-	-
3/4	5	-	1	3	1	-	-
5/6	5	1	-	3	1	-	+
7/8	5	-	1	3	1	-	+
9/10	5	1	-	3	-	1	-
11/12	5	-	1	3	-	1	-
13/14	5	1	-	3	-	1	+
15/16	5	-	1	3	-	1	+

Table 1: Composition of the mixed co-transfected plasmid DNA.
    a All the values indicate the amounts of plasmid DNA in µg.

5.2 Equipment, Chemicals and Solutions

5.2.1 Equipment

- a scanning system for β-radiometry (Bioscan Imaging Scanner, Packard)
- 20 x 20 cm TLC plates (silica gel)
- 5 µl capillary glass tubes
- a chromatography chamber for TLC
- 20 x 25 cm films for autoradiography (Kodak XAR-5)
- 20 x 25 cm exposure cassettes
- a rotary evaporator type Speed-Vac (Jouan RC1010)
- a microcentrifuge for Eppendorf tubes
- a thermostated water bath with racks
- a vortex mixer
- adjustable automatic pipettes with tips
- 1.5 ml Eppendorf tubes

5.2.2 Chemicals

- [^{14}C] chloramphenicol (25 µCi/ml, Amersham)
- acetyl CoA (25 mg, Sigma)
- ethyl acetate (200 ml)
- chloroform (2 L)

- methanol (500 ml)
- Tris (10 g)

5.2.3 Solutions

- 0.25 M Tris-HCl: add 3.02 g of Tris in 100 ml H_2O. Adjust the pH to 7.5 with HCl, filter, and autoclave for 20 min. Store at room temperature.
- 40 mM acetyl-CoA (32.4 mg/ml): add 778 µl H_2O to 25 mg of acetyl-CoA. Mix and store at -20°C

5.3 Experimental Procedure

The protocol is described for 1 sample.

Prepare dilutions for 17 samples (16 + 1 extra) by using an amount of cytosol equivalent to 100 units of β-galactosidase to perform CAT assay.

[1] In a 1.5 ml Eppendorf tube, add each cytosol extract (50-100 µl) and 0.25 M Tris-HCl pH 7.5 to a final volume of 120 µl.

[2] Add 0.1 µCi (4 µl) of [^{14}C] chloramphenicol that has been diluted to 40 µl final volume with 0.25 M Tris-HCl pH 7.5. Incubate at 37°C for 5 min.

[3] Add 40 µl of prewarmed (37°C) 4 mM acetyl CoA solution (1 in 10 dilution of stock solution in 0.25 M Tris-HCl pH 7.5). Mix, spin, and leave at 37°C for 1 h.

[4] Stop the reaction by extracting chloramphenicol and its derivatives with 900 µl ethyl acetate. Mix for 2 min. Spin for 1 min in a microfuge (12,000 rpm).

[5] Transfer the supernatant into another 1.5 ml Eppendorf tube. Avoid the interphase. Spin in the rotary evaporator with heat (40°C) for 30 min.

[6] Resuspend the pellet in 20 µl ethyl acetate. Spot all of the solution onto a TLC plate, using a 5 µl capillary tube. Place capillaries in all tubes, and spot sequentially 4 to 5 rounds to give more distinct spots, set equally at 2 cm from the bottom and 1 cm apart (marked with a pencil). Allow to dry for about 10 min at room temperature. Place the TLC plate in a saturated chromatography

chamber containing 200 ml of developing mixture (95% chloroform, 5% methanol) and with filter paper along the side of the chromatography chamber.

[7] When the migration line reaches 2 cm from the top, take the TLC plate out, and allow to dry at room temperature in a fume hood (5 min). Quantify using the β–scanner, and calculate the percentage of CAT acetylation.

[8] Expose to X-ray film at room temperature overnight (see 4.6 - Autoradiography).

5.4 Results

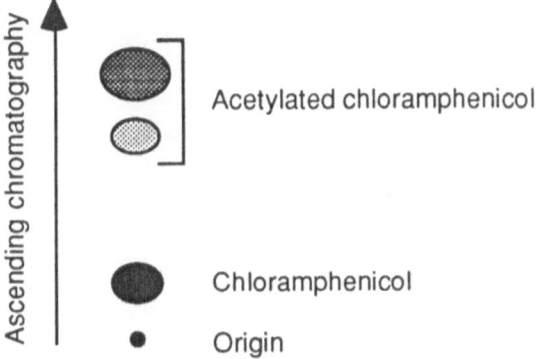

Figure 2: Schematic thin layer chromatogram obtained in the CAT assay.

References

1. Ausubel FM, Brent R, Kingston RE, Moore DD, Seidman JG, Smith JA, Struhl K (eds) (1990) Current Protocols in Molecular Biology. Wiley, New York
2. Gorman CM, Moffat LF, Howard BH (1982) Recombinant genomes wich express chloramphenicol acetyltransferase in mammalian cells. Mol Cell Biol 2:1044-1051
3. Green S, Issemann I, Scheer E (1988) A versatile *in vivo* and *in vitro* eukaryotic expression vector for protein engineering. Nucleic Acids Res 16:369

4. Issemann I, Green S (1990) Activation of a member of the steroid receptor superfamily by peroxisome proliferators. Nature 347:645-650

5. Tugwood JD, Issemann I, Anderson RG, Bundell K, McPheat W, Green S (1992) The mouse peroxisome proliferator activated receptor recognizes a response element in the 5' flanking sequence of the rat acyl-CoA oxidase gene. EMBO J 11:2:433-439

6 Synthesis of Oligonucleotides Used as Probes; Purification by HPLC

M. BENTÉJAC and M. BUGAUT

6.1 Introduction

Synthetic oligonucleotides are useful research tools in molecular biology [3, 7]. As hybridization probes, they can distinguish a perfect match in the target sequences from various single-base-pair mismatches [20, 21]. They have been used to introduce mutations in DNA at specific sites [8, 13, 15, 22, 23] as well as to probe for mutations at specific loci in genes [4, 9, 13, 20, 22, 23]. They have also been employed in the isolation of specific DNA sequences from a bank of cloned cDNA [6, 16, 19, 21] and in the determination of RNA secondary structure [11]. Moreover, the use of antisense DNA or RNA oligonucleotides to block the translation or processing of specific mRNAs provides a powerful tool for studies of the regulation of gene expression and development [2, 5].

Synthetic oligodeoxynucleotides are now widely available from on-site DNA synthesizers. The β-cyanoethyl phosphoramidite method of DNA synthesis, which is the current standard in the field, will be used in the present protocol. See ref [3] for details on the chemistry fundamentals.

In rodent liver, a number of peroxisomal proteins are inducible by peroxisomes proliferators [18]. The cloning and characterization of cDNA encoding many of these proteins were reported [10], making possible the study of their induction at the post-transcriptional level (Northern blotting, see 4.3) [1]. The synthetic oligonucleotide probes offer the possiblity to make nucleic acid hybridization assays without the need for plasmid isolation and purification to generate cDNA or RNA probes. In the experiment proposed in this session, we will synthesize short DNA oligonucleotides corresponding to the complementary fragments of the published rat cDNA sequences of acyl CoA synthetase [17], acyl CoA oxidase [12], and catalase [14]. Choice of the more suitable oligonucleotide sequences will be researched through DNA/Protein sequence computer analysis (Genbank and EMBL data library). Then, purification of these desired oligonucleotides will be performed by reversed-phase, high-performance liquid chromatography (HPLC) procedure.

6.2 Equipment and Reagents

6.2.1 Equipment

- a nucleic acid synthesizer (Cyclone Plus, Milligen/Biosearch)
- a liquid chromatography system (Waters 625)
- a 3.9 x 150 mm column packed with 5 μm, 300Å C18 particules (Delta-Pak, Waters)
- a tunable absorbance detector (Waters 486)
- a flatbed recorder (Servogor 120, Waters)
- a vaccum filter for HPLC solvents (Millipore)
- a rotary evaporator type Speed-Vac (Jouan RC 1010)
- a spectrophotometer (Kontron series Uvikon)
- a 1 ml and a 20 ml sterile Luer tip syringes
- a 25 μl syringe (Hamilton)
- 0.45 μm filters (Millipore FH)
- 0.45 μm filters (Millipore HA)
- adjustable automatic pipettes (Gilson or Labsystems) and tips
- 1.8 ml tubes with screw cap (Nunc)
- 1.5 ml Eppendorf tubes

6.2.2 Reagents

6.2.2.1 *Solutions for β-Cyanoethyl Phosphoramidite Chemistry (Waters)*

- Amidite Diluent (acetonitrile) (100 ml)
- Amidite Activator solution (tetrazole) (120 ml)
- Cap A solution (acetic anhydride) (90 ml)
- Cap B solution (N-methylimidazole) (125 ml)
- Deblock solution (dichloroacetic acid) (450 ml)
- Oxidizer solution (iodine), 150 ml

6.2.2.2 *β-Cyanoethyl Phosphoramidite Monomers for DNA Synthesis (0.5 g of each, Waters)*

The aramidites are protected at the 5'-OH position with dimethoxytrityl (DMT) groups.
- DMT-dAdenosine (N^6-benzoyl) cyanoethyl phosphoramidite
- DMT-dCytidine (N^4-benzoyl) cyanoethyl phosphoramidite
- DMT-dGuanosine (N^2-isobutyryl) cyanoethyl phosphoramidite
- DMT-Thymidine cyanoethyl phosphoramidite

6.2.2.3 Prepacked, Disposable Reaction Columns for DNA Synthesis (Waters)

The columns contain derivatized, controlled 500Å pore glass support for a synthesis scale of 0.2 µmol.
- DMT-dAdenosine (N^6-benzoyl) column
- DMT-dCytidine (N^4-benzoyl) column
- DMT-dGuanosine(N^2-isobutyryl) column
- DMT-Thymidine column

6.2.2.4 Other Reagents

- triethylamine (99.5 %, Aldrich)
- 30% ammonium hydroxide (Aldrich)
- DNA synthesis grade acetonitrile (water content less than 0.005 %, Waters)
- HPLC grade acetonitrile (Aldrich)
- Glacial acetic acid
- Milli-Q water

Millipore is a registered trademark of Millipore Corporation.
Milli-Q is a registered trademark of Millipore Corporation.
Cyclone, Delta Pak, Milligen/Biosearch, Waters are trademarks of Millipore Corporation.
Gilson is a registered trademark of Gilson Medical Electronics, Inc.
Hamilton is a registered trademark of Hamilton Co.

6.4 Experimental Procedure

6.4.1 Operating Procedure to Perform a Synthesis

6.4.1.1 Flow Rate Calibration

Reagent flow rates are dependent on the gas pressure used to drive the solutions to the column. The gas pressure is determined by the setting of the internal regulator. Flow rates are increased or decreased by increasing or decreasing the regulator setting.

The helium pressure has to be set to the starting pressure appropriate to the synthesis scale (5.7 psi with 0.2 µmol scale). When performing calibration, you should measure a characteristic flow rate for each valve train (A-train and B-train) to ensure that proper pressure has been established and that both trains are flowing freely.

The C monomer reservoir flow rate should be 3.0 ml/min. Measure this flow rate precisely (tests B-train):

[1] Place a disposable column corresponding to the scale of the synthesis to be run.

[2] Fill the acetonitrile wash bottle with dry acetonitrile and place about 20 ml of dry acetonitrile in the C monomer reservoir. All other reservoirs should be in place.

[3] Connect the column wash line to a 10 ml graduated cylinder to collect the flow volumes.

[4] Select "manual" from the main menu screen. Select "wash" from the manual screen.

[5] Use the up and down keys to select the C monomer function (reservoir 7). Press the "on" key and collect the acetonitrile for 1 min. Press the "off" key after the minute has elapsed.

[6] Adjust the internal pressure regulator to increase or decrease the flow rate. The excess pressure requires venting when adjusting from a higher to lower pressure. Turn the "system pressure" switch to its off position, and briefly unscrew the external acetonitrile wash bottle, the C reservoir, and all of the ancillary reagent bottles.

[7] Repressurize by turning the "system pressure" switch back on. Allow the flowmeter to settle.

[8] Repeat steps 3 to 7 until the desired flow rate is obtained.

[9] After calibrating the C monomer reservoir precisely, check the wash A flow rate. Measure the wash A flow rate (tests A-train); this flow rate should be 4.4 ml/min (±10%).

[10] Select "manual" from the main menu screen. Select "wash" from the manual screen.

[11] Use the up and down keys to select the wash A function (reservoir 11). Press the "on" key and collect the acetonitrile for 1 min. Press the "off" key after the minute has elapsed.

[12] Measure the volume in the cylinder and calculate the flow rate. The flow rate should be within 10% of its nominal value without readjusting the helium pressure.

6.4.1.2 Performing a Synthesis

[1] Add 20 ml of Amidite Diluent to each bottle of amidite and attach the amidite bottles to the numbered positions (*i.e.* A to position 6, C to position 7, G to position 8, and T to position 9). A bottle containing 20 ml of dry acetonitrile must be attached to position 10 (reservoir X) because it will not be used during synthesis. The amidites must remain anhydrous; therefore, work quickly to minimize the exposure to air.

[2] Attach other reagents to their respective positions (Amidite Activator to position 1, Cap B to position 2, Cap A to position 3, Oxidizer to position 4, Deblock to position 5). Attach a 4 L bottle of dry, nucleic acid synthesis grade acetonitrile to the external wash bottle cap assembly.

[3] Pressurize the instrument by turning the "system pressure" switch ON and watch the gas flowmeter to check for leaks.

[4] Attach a disposable reaction column containing the appropriate initial support-bound protected nucleoside corresponding to the 3' end nucleotide of the oligonucleotide sequence that must be synthesized.

[5] Select the "prime" mode from the main menu. Select "prime 1". This option samples all of the reagent and monomer reservoirs to ensure that the delivery lines are filled with fresh reagents. After sampling the reservoirs, flush the lines with acetonitrile in preparation for synthesis.

[6] Enter the DNA sequence to be synthesized in the 5' to 3' direction from the "seq" mode of the main menu. The A, C, G and T keys are used to enter the standard nucleotides. To enter degenerate sites, press the "mixed site key".

[7] Select the "run" mode from the main menu. Indicate whether the last 5' DMT group should be removed or retained. Retain the DMT group as purification will be by reverse phase HPLC method.

[8] Before you begin synthesis, select the "print" option. On the connected printer, the print softkey will allow the sequence, the sequence length, the base frequencies, and the purine/pyrimidine ratio to be printed. Verify that the correct sequence has been entered. Start the synthesis.

6.4.1.3 Cleaving/Deprotecting

Once a synthesis is complete, the oligonucleotide must be cleaved from the support and fully deprotected:

[1] Remove the support from the column by puncturing the reaction column and pouring out the support into an Eppendorf tube. Add 1.5 ml of 30% ammonium hydroxide. Let stand at room temperature for 120 min to cleave the oligonucleotide from the support. The ammonium hydroxide must be fresh. To ensure a concentrated solution, stock the ammonium hydroxide at 4°C after opening and keep tightly capped.

[2] Decant the DNA-containing supernatant into a small screw cap vial (Nunc). Ensure that the is well sealed.

[3] Deprotection is accomplished by heating the tube at 55°C overnight. If the oligonucleotide is purified by reversed-phase HPLC method, the highest yield of product is obtained by letting the reaction proceed at room temperature for 24 hours (this minimizes any detritylation caused by heat).

[4] Cool the tube and transfer the DNA into an Eppendorf tube suitable for rotary evaporation. Evaporate the solution to dryness (with the heat off for 5'-DMT containing oligonucleotides). A pellet should be visible.

6.4.1.4 Determining Yield

The yield of crude or purified oligonucleotide is determined by spectrophotometry:

[1] Dissolve the pellet of crude or purified oligonucleotide in 1 ml of Milli-Q water.

[2] Dilute 10 μl of this stock solution in 1 ml of water.

[3] Read the absorbance at 260 nm (the dilution should bring the absorbance readings into the linear range of the spectrophotometer).

[4] Calculate the total OD_{260} units of oligonucleotide obtained for a 0.2 μmol synthesis as follows: Total OD_{260} = Absorbance x 100. One OD_{260} unit is equivalent to approximatively 33 μg/ml of oligonucleotide (e.g. 0.01 μmol/ml of a 10-mer oligonucleotide).

6.4.2 Purification by Reversed-Phase HPLC

Purification of the desired oligonucleotide from a crude mixture involves separation of the oligomer from the short failure sequences and the by-products of deprotection. Reversed-phase HPLC procedure requires about 40 min. The protecting groups and other by-products from the synthesis elute at about 5-10 min The failure sequences with the free 5'-hydroxyl groups will elute at 20-25 min.

The DMT-protected oligonucleotide peak elutes at about 30 min depending on sequence length. After detritylation, the oligonucleotide peak will run at about 20-25 min depending on length. Purification by reversed-phase HPLC can be performed as a four step process. The first step is to perform an analytical run as a preliminary evaluation of the synthesis. A preparative run is performed to isolate a large quantity of the tritylated oligonucleotide (50-100 OD_{260} units). The oligonucleotide is then detritylated. A final HPLC run will determine if all of the oligomer was detritylated and if adequate purification was achieved.

6.4.2.1 Instrument Settings

- Flow rate   0.8-1.0 ml/min
- Wavelength   260 nm
- Sensitivity   0.2-2.0 AUFS
- Temperature   30°C
- Injection Volume   20 µl
- Run time   40 min
- Equilibration delay   20 min
- Time constant   1.0
- Silk option   ON
- Reservoirs to sparge   A and B
- High pression limit   1500 psi
- Chart   10 mV full scale output, plot %B
                               to a trace gradient curve profile during analysis.

Select the following data which have to appear on the screen:

Gradient Table

Time	Flow	%A	%B	Curve
Initial	1.0	95	5	*
40.0	1.0	60	40	6 (linear gradient)
41.0	1.0	95	5	11 (step gradient)

Events Table

Time	Event	Action
Initial	Temp	30°C
0.00	SPRG	100
10	SPRG	10

6.4.2.2 Preparing HPLC Eluents

Solvents such as acetonitrile must be HPLC grade. Solutions must be filtered to remove microparticulates above 0.45 μm.

- Eluent A: 100 mM triethylammonium acetate (TEAA), pH 6.5, prepared as follows:

  - Mix 3 ml of glacial acetic acid and 800 ml of milli-Q water.
  - Add 14.0 ml of triethylamine and stir.
  - Bring the pH to 6.5 with the dropwise addition of glacial acetic acid.
  - Dilute to 1 L with Milli-Q water.
  - Vacuum filter the solution through a 0.45 μm filter (Millipore HA).

- Eluent B: 95:5 (v/v) acetonitrile / Milli-Q water. Vacuum filter the solution through a 0.45 μm filter (Millipore FH)
- 80% (v/v) acetic acid in Milli-Q water
- 1% (v/v) triethylamine in Milli-Q water

2.4.2.3 Preparing and Injecting Samples

Analytical Run
[1] Dissolve the DNA pellet (from a 0.2 μmol synthesis) in 200 μl of eluent A.

[2] Centrifuge briefly to remove undissolved silicates and transfer the supernatant.

[3] Dilute 20 μl of supernatant to 100 μl. Inject 10 μl of the diluted solution (you should inject about 0.5 OD_{260} unit of crude oligonucleotide).

Preparative Run
- Load crude oligonucleotide from a 0.2 μmol synthesis onto the column.

6.4.2.4 Detritylating Purified Product

[1] Using a rotary evaporator, concentrate to dryness the fraction containing the DMT-protected purified product.

[2] Resuspend the residue in 100 μl of 80% acetic acid. After 20 min at room temperature, dry down again.

[3] Resuspend in 0.5 ml of 1% triethylamine and concentrate to dryness. This last step minimizes the exposure of the oligonucleotide to acid.

[4] Redissolve in eluent A to perform another analytical run.

[5] Evaporate an aliquot from step 4 and dissolve in Milli-Q water to determine the yield in OD_{260} units.

[6] Store the oligonucleotide as a solid at -20°C.

Acknowledgements

The authors wish to thank Millipore Corporation for the permission to incorporate parts of their "Milligen/Biosearch Cyclone Plus operator's manual" in this manuscript.

References

1. Causeret C, Bentéjac M, Bugaut M (1993) Proteins and enzymes of the peroxisomal membrane in mammals. Biol Cell 77:89-104
2. Cohen JS (1991) Antisense oligodeoxynucleotides as antiviral agents. Antiviral Res 16:121-133
3. Eckstein F (1991) Oligonucleotides and analogues: A practical approach, IRL Press, Oxford
4. Farr CJ, Saiki RK, Erlich HA, McCormick F, Marshall CJ (1988) Analysis of RAS gene mutations in acute myeloid leukemia by polymerase chain reaction and oligonucleotide probes. Proc Natl Acad Sci USA 85:1629-1633
5. Hélène C, Toulmé JJ (1990) Specific regulation of gene expression by antisense, sense and antigene nucleic acids. Biochim Biophys Acta 1049:99-125
6. Heller RA, Song K, Freire-Moar J (1992) Rapid screening of libraries with radiolabeled DNA sequences generated by PCR using highly degenerate oligonucleotide mixtures. Biotechniques 12:30-35
7. Itakura K, Rossi JJ, Wallace RB (1984) Synthesis and use of synthetic oligonucleotides. Annu Rev Biochem 53:323-356

8. Ito W, Ishiguro H, Kurosawa Y (1991) A general method for introducing a series of mutations into cloned DNA using the polymerase chain reaction. Gene 102:67-70
9. Karnik SS, Khorana GH (1990) Assembly of functional rhodopsin requires a disulfure bond between cysteine residues 110 and 187. J Biol Chem 265:17520-17524
10. Latruffe N (1992) Les peroxysomes et la proliferation cellulaire ou la prise en considération d'un organite méconnu. Médecine/Sciences 8:239-248
11. McWilliams RA, Glitz DG (1991) Localization of a segment of 16S RNA on the surface of the small ribosomal subunit by immune electron microscopy of complementary oligodeoxynucleotides. Biochimie 73:911-918
12. Miyazawa S, Hayashi H, Hijikata M, Ishii N, Furuta S, Kagamiyamla H, Osumi T, Hashimoto T (1987) Complete nucleotide sequence of cDNA and predicted aminoacide sequence of rat acyl-CoA oxidase. J Biol Chem 262:8131-8137
13. Norrander J, Kempe T, Messing J (1983) Construction of improved M13 vectors using oligodeoxynucleotide-directed mutagenesis. Gene 26:101-106
14. Osumi T, Ozasa H, Miyazawa S, Hashimoto T (1984) Molecular cloning of cDNA for rat liver catalase. Biochem Biophys Res Commun 122:831-837
15. Sayers JR, Krekel C, Eckstein F (1992) Rapid high-efficiency site-directed mutagenesis by the phosphorothioate approach. Biotechniques 13:592-596
16. Suggs SV, Wallace BR, Hirose T, Kawashima EH, Itakura K (1981) Use of synthetic oligonucleotides as hybridization probes: isolation of cloned cDNA sequences for human β^2-microglobulin. Proc Natl Acad Sci USA 78:6613-6617
17. Suzuki H, Kawarabayasi Y, Kondo J, Abe T, Nishikawa K, Kimura S, Hashimoto T, Yamamoto T (1990) Structure and regulation of rat long-chain acyl-CoA synthetase. J Biol Chem 265:8681-8685
18. Van Den Bosch H, Schutgens RBH, Wanders RJA, Tager JM (1992) Biochemistry of peroxisomes. Annu Rev Biochem 61:157-197
19. Wallace RB, Johnson MJ, Hirose T, Miyake T, Kawashima EH, Itakura K (1981) The use of synthetic oligonucleotides as hybridization probes. II. Hybridization of oligonucleotides of mixed sequence to rabbit β-globin DNA. Nucl Acids Res 9:879-894
20. Wallace RB, Shaffer J, Bonner J, Hirose T, Itakura K (1979) Hybridization of synthetic oligodeoxyribonucleotides to $\phi\chi$ 174 DNA: the effect of single base pair mismatch. Nucl Acids Res 6:3543-3557
21. Wood WI, Gitschier J, Lasky LA, Lawn RM (1985) Base composition-independent hybridization in tetramethylammonium chloride: A method for oligonucleotide screening of highly complex gene libraries. Proc Natl Acad Sci USA 82:1585-1588
22. Zoller MJ, Smith M (1982) Oligonucleotide-directed mutagenesis using M13-derived vectors: an efficient and general procedure for the production of point mutations in any fragment of DNA. Nucl Acids Res 10:6487-6500
23. Zoller MJ, Smith M (1983) Oligonucleotide-directed mutagenesis of DNA fragments cloned into M13 vectors. In Wu R, Grossman L and Moldave K (eds). Methods in Enzymology, Vol. 100, Academic, New York, pp. 468-500

7 Computer Analysis of DNA and Protein Sequences

M.C. CLEMENCET and N. LATRUFFE

7.1 Introduction and Aims

Data banks are tremendously powerful tools for molecular biologists and are becoming essential to elucidate the chemical basics of life and evolution. The CITI2 (Centre Inter Universitaire de Traitement de l'Information) in Paris allows, through the BISANCE server, interaction with the main data banks, the EMBL data library at Heidelberg, Germany, GenBank in Cambridge, MA, USA, or NBRF-PIR (Protein Identification Response) of Georgetown University, Washington, D.C., USA [1].

Many possibilities are offered, such as access to different data banks to choose and work with protein or nucleic acid sequences, DNA sequencing aid, treatment of sequences, access to primary and secondary structures, search for motifs, comparison of sequences together and with the bank, search for homology, transcription, translation, protein structure, genetic and restriction maps, DNA or protein phylogenetic trees, etc...

From 150 different programs, we propose using some of those applied to some proteins, especially proteins related to peroxisomes and to their encoding genes.

7.2 Outline

Part A - Data Bank Access

Search A_1 : EMBL Data Library - Peroxisomal Acyl CoA Oxidase
Search A_2 : GenBank - Peroxisome Proliferator Activated Receptor (PPAR)
Search A_3 : NBRF-PIR - Peroxisome Proliferator Activated Receptor (PPAR)
Search A_4 : PROSITE DATA File - Peroxisomal Targeting Sequences (PTS)

Part B - Translation of Nucleic Acid Sequences into Protein Sequences and Saving

Part C - Alignments of Protein Sequences

Search C₁ : Alignments of Two Sequences
Search C₂ : Alignments of Several Sequences

Search C_1 : Alignments of Two Sequences
Search C_2 : Alignments of Several Sequences

Part D - Primary Structure of Nucleic Sequences

Search D_1 : Base and Codon Frequencies
Search D_2 : Consensus Sequences
Search D_3 : Repeating Sequences

Part E - Protein Data

Search E_1 : Information, Composition and Sequence of a Protein
Search E_2 : Molecular Mass of a Protein
Search E_3 : Secondary Structure of a Protein
Search E_4 : Minimum Degenerescence Probes

7.3 Procedure

7.3.1 First Operation

Connect to the BISANCE server by Transpac line, using a modem and a Macintosh unit.

CITI2	B I S A N C E	Vax 6410/8530

MAIN MENU

Menu 1. Access and data manipulations
Menu 2. Sequencing programmes
Menu 3. Primary structure of sequences
Menu 4. Secondary structure of sequences
Menu 5. Sequence comparison
Menu 6. Restriction enzymes and maps
Menu 7. Sequence translation : nucleic acid<-> protein sequences
Menu 8. Access and protein data analysis
Menu 9. Search in data banks and prerecorded programmes

Menu 10. Useful programmes and library data banks (references, Genatlas)
Menu 11. Phylogenetic trees construction (nucleic acids and proteins)
Menu 12. Packages or complementary programmes

7.3.2   Part A - Data Bank Access

7.3.2.1 Search A_1: EMBL Data Library - Peroxisomal Acyl CoA Oxidase

By using the EMBL data library of nucleic acids, find information on peroxisomal acyl CoA oxidase (Menu 1-1).

OPTION (Help = ?)
1 - 1
1- Interrogation of the banks (GenBank, EMBL, NBRF, SWISSPROT)
Bank to use : GENBANK (G) EMBL (E) NBRF (N) SWISSPROT (S)
E

EMBL Data Base Content
EMBL Library Release 31 + daily update (June 1992)
103,217,018 bases; 81,084 sequences; 70,762 subsequences; 45,132 references.
M. Gouy & M. Jacobzone, Laboratory of Biometry, University of Lyon 1

Command? (H for command list)
K
(Give part(s) of word(s) you want be present in keywords, ex: RIB TEIN)
keyword searched? (or ALL for all keywords or H for help)
coenzyme A oxidase

0/ ACYL-COENZYME A OXIDASE
Command? (H for command list)
S E
Enter your selection criteria, or H(elp) (EX: sp=homo sapiens et k=globinà)
k=acyl-coenzyme a oxidase

Sequence list named LIST1   contains   7 seqs
Command? (H for command list)
S H
List name, sequence, or accession #, or H(elp)? âdefault=LIST1
Short description of list: LIST1

CMPOX4   Candida maltosa POX4 gene for peroxiomal acyl CoA oxidase
          (AOX, EC 1.3.99.3)   =2975 bp

93

CTPOX4   Yeast (C.tropicalis) POX4 gene encoding acyl-coenzyme A oxidase II
          (PXP-4), complete cds.   =2845 bp
CTPOX42   Candida tropicalis mRNA fragment for acyl-coenzyme A oxidase
CTPOX5   Yeast (C.tropicalis) POX5 gene encoding acyl-coenzyme A oxidase I
          (PXP-5), complete cds.   =2624 bp
RNACOA1   Rat acyl-coA oxidase mRNA, complete cds.   =3741 bp
RNACOA2   Rat acyl-CoA oxidase gene, 5' end.   =1558 bp
SCPOX1   Yeast (S.cerevisiae) acyl-coenzyme A oxidase (POX1) gene, complete
          cds.   =2915 bp

Command? (H for command list)
IN
List name, sequence, or accession #, or H(elp)? âdefault=LIST1ê
RNACOA1

Choose one or several of the following topics, finish with two <RETURN>s
AC DT KW OS OC OG RN RC RP RA RT RL DR CC FH FT SQ SEQ ALL
H(elp)
all
?
ID   RNACOA1   standard; RNA; ROD; 3741 BP.
AC   J02752;
DT   06-JUL-1989 (Rel. 20, Last updated, Version 1)
DT   02-APR-1988 (Rel. 15, Created)
DE   Rat acyl-coA oxidase mRNA, complete cds.
KW   acyl-coenzyme A oxidase; oxidase.
OS   Rattus norvegicus (rat)
OC   Eukaryota; Animalia; Metazoa; Chordata; Vertebrata; Mammalia;
OC   Theria; Eutheria; Rodentia; Myomorpha; Muridae; Murinae.
RN   â1ê
RP   1-3741
RA   Miyazawa S., Hayashi H., Hijikata M., Ishii N., Furuta S.,
RA   Kagamiyama H., Osumi T., Hashimoto T.;
RT   "Complete nucleotide sequence of cDNA and predicted amino acid
RT   sequence of rat acyl-CoA oxidase";
RL   J. Biol. Chem. 262:8131-8137(1987).
DR   SWISS-PROT; P07872; CAO1$RAT.
DR   SWISS-PROT; P11354; CAO2$RAT.
CC   Draft entry and computer-readable sequence of â1ê kindly provided
CC   by T.Osumi, 02-APR-1987. Two species of acyl-CoA oxidase mRNA were
CC   isolated. They differed only in nulceotides between 270 and 429,
CC   that is residues 90-143. The authors proposed that the enzyme
CC   acyl-CoA oxidase contains three components A,B and C, " the latter
CC   two being produced from the first by a proteolytic cleavage.

FH   Key   Location/Qualifiers
FH
FT .PE1 CDS   74..2059
FT   /note="acyl-CoA oxidase (E.C 1.3.3.6)"
SQ   Sequence 3741 BP; 981 A; 908 C; 894 G; 958 T; 0 other;

Command? (H for command list)
IN
List name, sequence, or accession #, or H(elp)? âdefault=RNACOA1
Choose one or several of the following topics, finish with two <RETURN>s
AC DT KW OS OC OG RN RC RP RA RT RL DR CC FH FT SQ SEQ ALL
H(elp)
seq
?
ID   **RNACOA1   standard; RNA; ROD; 3741 BP.**
DE   **Rat acyl-coA oxidase mRNA, complete cds.**

```
   1 acacgggtcg ttgctttggt gtctgtcact tctgtcgcca cctcctctgc caacaccaac
  61 actgacctcc gtcatgaacc ccgacctgcg caaggagcgg gcctccgcca ccttcaatcc
 121 ggagttgatc acgcacatct tggatggcag tccggagaat accggcgcgcc gtcgagaaat
 181 tgagaacttg attctgaacg acccagactt ccagcatgag gactataact tcctcactcg
 241 aagccagcgt tatgaggtgg ctgttaagaa gagtgccacc atggtgaaga agatgagcga
 301 atatggcatc tcggaccctg aagaaatcat gtggtttaaa aaactatatt tggccaattt
 361 tgtggaacct gttggcctca attactccat gtttattcct accttgctga atcagggcac
 421 cactgctcag caggagaaat ggatgcgccc gtcccaagaa ctccagataa ttggcaccta
 481 cgcccagacg gagatgggcc acggaactca tcttcgaggc ttgcaaacca ctgccacata
 541 tgaccccaag acccaagagt tcattctcaa cagcccctact gtgacttcca ttaagtggtg
 601 gcctggggga cttgggaaaa cttccaatca cgcaatagtt ctggctcagc tcatcactca
 661 aggagagtgc tacgggttac atgcctttgt tgtccctatc cgtgaaattg ggacccacaa
 721 gcccttgcca ggcatcactg tcggggatat cggtcccaaa tttggttatg aagagatgga
 781 taacggctac ctgaagatgg acaattaccg tattcccaga gagaacatgc tgatgaaata
 841 cgcccaggtg aagcctgatg gcacatatgt aaagcctttg agtaacaagc tgacgtatgg
 901 gaccatggtt tttgtgaggt ccttcctcgt gggaaatgca gctcagagtc tgtccaaggc
 961 ttgcacaatc gccatacgat acagcgctgt gaggcgccag tctgaaatca agcaaagcga
1021 accagaacca cagatttttgg attttcaaac ccagcagtat aaactcttcc cgctcctggc
1081 cactgcctat gccttccact tcgtaggaag gtacatgaag gagacctacc ttcgaattaa
1141 tgagagcatt ggccaagggg acctgagtga actgcctgag cttcacgccc tcactgctgg
1201 gctgaaggct tttactactt ggacagccaa tgctggcatc gaagaatgtc gaatggcctg
1261 cggcgggcac ggctattctc acagcagtgg gattccaaat atttacgtca cttttacccc
1321 ggcctgcacc ttcgagggag agaacactgt tatgatgctg cagacagcca ggttcttgat
1381 gaaaatctat gaccaggtgc ggtcgggaa gttggtgggt ggtatggtgt catacctgaa
1441 tgacctgccg agtcagcgga tccagccaca gcaggtggca gtctggccaa ctatggtgga
1501 catcaacagc ctggagggcc tgacagaagc ctacaagctt cgtgcagcca gattggtaga
1561 aatcgctgca aaaaacctttc agactcacgt gagtcacagg aagagcaagg aagtagcatg
1621 gaacctaacc tctgtcgacc ttgttcgggc aagtgagggcg cattgccact acgtggtcgt
```

95

```
1681 taaggtcttc tcagacaaac tccccaagat tcaagacaaa gccgtccaag ctgtgctgag
1741 gaacctgtgt ctcttgtatt ctctctatgg gatcagccag aaaggagggg actttcttga
1801 ggggagcatc atcacagggg ctcagctgtc acaagtaaac gctcggatcc tggagctgct
1861 caccctgatc cgccccaatg ctgttgctct ggtggatgcc tttgacttta aggacatgac
1921 acttggctct gttcttggcc gctatgatgg aaatgtgtat gaaaacttgt ttgagtgggc
1981 caagaaatcc ccactgaaca aaacagaggt ccatgaatct taccacaagc acttgaagcc
2041 cctgcagtcc aagctttgaa gtttccctgg gacacgtctg agctccacaa gcagcagaaa
2101 ctctctcctc tactcactaa tccttgtgaa atcgtcatca aatttgtgta gctacagagc
2161 aaatgatggg tttcttttcc tccctataag taaagagaaa tgaacagact ttagagatta
2221 aatgagaaat tcggtgttgt aagtgcagta atgcagacag agacgtagga actcagaaag
2281 cttcagagct gtggtctgac ttgcatttgt actgctgcta atctcagtag gccttgactc
2341 tggagaatta acagagtttt aactacaaat acttacttat tttcacattt ttactgctaa
2401 tcactggata tatgtttttt aaacaaaagt gttatataga gtggaatttt ccaggcattc
2461 gtgcctacca ctttctgatc tactgctaag agcaggagtt tgggggccag aaactaatag
2521 aaaccttgat gtgggtgtgt ggctgtcaca caggccgctg ctgcctgcca tgtgggtgtg
2581 tcacccctat taactgtcac attaacatag tcaacaagag gactccttca acacccaccc
2641 accaagaaac cagagttccc gatgggcaca agcacccact ccaggctcag ccttcgtggt
2701 agtggcacta ctgtgccttt gacccactt tttgacacag ttaagttact tgtcttacct
2761 ccaggctttc agccattgcc tggatttcag tcatggtggc tgaccttccc tttcttgctt
2821 gccttccttc tcctgaaaga gataaaaagag atagtagtca gcctctcctc atagattaag
2881 tatatggaga gccctcagct atggtattac tgtattttgg tgactttgtt aagtaaattt
2941 cctgggacaa tcctgatttg aaagattctg tattctgttg cacaagctat taaaatgctc
3001 agtggtcacc aaagtatttc acacacacac actcacaaat cactcaggtc accttttaca
3061 ctagaaataa caggaaaagcc ctggccacag ccatctgctg agagtgttag ttgagatgtt
3121 tcgttgaaga tggctgaaga aagtgcaggc cctggggagt ttctccttgc tgagagccgg
3181 ctctgtggtg agcccctagc agccttacga ggcggtgaaa cggccccttc agatgggagc
3241 agcctctaca atcattctga gcttaaaggt gaaatatgca cctttgtcc tataaatgtc
3301 ctataaatgc ttgggggggag gggtaatgtt ttgtttttt tttcttgaaa cagagcccag
3361 aatggccttg aactctgatc ctcacagcct ggtaaatgct gggcctacag gtgtgcgccg
3421 acatgtctgg ctgatcggag ctctttattc ttaaaagcac agtagggaga tgattgtaac
3481 attaagtctg tgtctgtggc attcgcattg tgagagcagt tccttagaac agttctgaga
3541 gcacagcatc aatgtagtga aaacgatgtc gacgcctgac atgaaatcac aacccctcggg
3601 gtcatcgaca caagccttaa cctttctcct ggaatgactg gtaatcccctt aagattgaca
3661 gtacacagca tgtcaccttg ttggggtttc tattgacagt aattcatatt ctggaaagga
3721 actaataaat ataaatgacc g
```

//

Command? (H for command list)

IN

List name, sequence, or accession #, or H(elp)?

rnacoa2

Choose one or several of the following topics, finish with two <RETURN>s
AC DT KW OS OC OG RN RC RP RA RT RL DR CC FH FT SQ SEQ ALL
?
all

ID   **RNACOA2**   **standard; DNA; ROD; 1558 BP.**
AC   J02753;
DT   06-JUL-1989 (Rel. 20, Last updated, Version 1)
DT   02-APR-1988 (Rel. 15, Created)
DE   **Rat acyl-CoA oxidase gene, 5' end.**
KW   acyl-coenzyme A oxidase; oxidase.
OS   Rattus norvegicus (rat)
OC   Eukaryota; Animalia; Metazoa; Chordata; Vertebrata; Mammalia;
OC   Theria; Eutheria; Rodentia; Myomorpha; Muridae; Murinae.
RN   âlê
RP   1-1558
RA   Osumi T., Ishii N., Miyazawa S., Hashimoto T.;
RT   "Isolation and structural characterization of the rat acyl-CoA
RT   oxidase gene";
RL   J. Biol. Chem. 262:8138-8143(1987).
DR   SWISS-PROT; P07872; CAO1$RAT.
DR   SWISS-PROT; P11354; CAO2$RAT.
CC   Draft entry and computer-readable sequence of âlê kindly provided
CC   by T.Osumi, 02-APR-1987. Six "GC" box sequences were found at
CC   positions 795-800, 926-931, 1141-1146, 1192-1197, 1197-1202 and
CC   1216-1221. A middle-interspersed repetitive or identified (ID)
CC   sequence was observed at nucleotides 390-482. This sequence was
CC   flanked on either sides by 16-bp direct repeats at positions
CC   374-389 and 483-498
FH   Key   Location/Qualifiers
FH
FT .PE1 CDS   1349..1457
FT   /note="acyl-CoA oxidase, exon 1 (E.C 1.3.3.6)"
FT   intron   1458..>1558
FT   /note="aCoA intron A"
SQ   Sequence 1558 BP; 378 A; 421 C; 378 G; 381 T; 0 other;

Command? (H for command list)
IN
List name, sequence, or accession #, or H(elp)? âdefault=RNACOA2ê
Choose one or several of the following topics, finish with two <RETURN>s
AC DT KW OS OC OG RN RC RP RA RT RL DR CC FH FT SQ SEQ ALL ?
seq

ID RNACOA2   standard; DNA; ROD; 1558 BP.
DE   **Rat acyl-CoA oxidase gene, 5' end.**

       1 aagctttggc taagaggaat aaaattcaca ataatcaaac taaacaaccc atcaatgcct

      61 aagagccctt gaaaatatcc ccagtagaac cttgttcagg catatggagc ctgagtcaaa

     121 ggtacagtca acaatgattt ggtctttatt taactcttca gaattttcct cattttccct

```
 181 tttgaactta atacattttg tggagctgag gatatctcct gcatcaagaa aaaagtatat
 241 taaaatattg aatttttgga gctcccttaa tttttgtact caagataaag ccttcattta
 301 ttttacctgt tctcagcact gttaatgtaa aagggtctta catacatcgt tacaagggca
 361 agtttagagtg gtaaagaaat gtagttcttg gggctgggga tttagctcag tggtagagtg
 421 cttacctagg aagcgcaagg ccctgggttc ggtccccagc tcccaaaaaa aaaaaaaaaa
 481 aaaagaaatg tagttcttct ctggggcatg aaatcaaatg aagaaagagt gcattttact
 541 tagcacctaa aggctttttt ttttttaaata tttatttta tgtgtacact atagccatcc
 601 acagacacac cagaagaggg tatcagatcc cattacagag ggttgtgagc cccatgtgg
 661 ttgctgggaa ttgaactcag gacaatgaac cctttcccga acgtgacctt tgtcctggtc
 721 ccctttttgct cctatttggt taggttccag taaaatcaac agttccagta aattcccta
 781 ttagcctctt cactccgccc gagaccctga acacccactc cctctggtt tcccaggact
 841 gtttagcgaa gagactctcc ccgaccgcac cccagctcca tcgcccaagc ctcaaagtag
 901 gtgagcccca cagcgcaggc gcaatcccgc ccctcaagac cctgaccaat ccgttgcgcg
 961 agtctcctgg gctccttttc gcccggaaag atcacgtgaa cctggtgaag tgacgtctgg
1021 aaagagccct actcccct tc cagtgccacc ggggtttctt gaccttctac cttgctctcc
1081 ttggcgagga gccgctcagg gacctcccct ggcccgcagc aggggattcg tccgtccggc
1141 gggcggtacc agcgtccgcg attgcagccg gtcacgtgat gagtggcggc agggcggggc
1201 ggggcgggcc agagtgggcg gggccgagga ctgtctagct ccgctcttag ggccctgtcg
1261 gcgcctgggc agcggacacg ggtcgttgct ttggtgtctg tcacttctgt cgccacctcc
1321 tctgccaaca ccaacactga cctccgtcat gaacccgac ctgcgcaacgg agcgggcctc
1381 cgccaccttc aatccggagt tgatcacgca catcttggat ggcagtccgg agaataccog
1441 gcgccgtcga gaaattggtg agaggagaat cggtggaggg ccaggcccct ctcctacga
1501 acgagaggaa aggacccccc ccccgtgaa tgaggatttg gggcctctgg agtttagg
```

7.3.2.2 Search A₂ : GenBank - Peroxisome Proliferator Activated Receptor (PPAR)

By asking GenBank, the nucleic acid data bank, find information on the PPAR.

GenBank Version 75 (15 February 1993)
126,124,047 bases; 106,637 sequences; 71,658 subsequences; 49,782 references.
M. Gouy & M. Jacobzone, Laboratory of Biometry, University of Lyon I

Sequence list named LIST1   contains   11 seqs
Short description of list: LIST1
HUMPPARA   Human peroxisome proliferator activated receptor mRNA,
       complete cds   =3301 bp
HUMPPARA.NUCI   Location/Qualifiers   (length= 1326 bp)
       CDS   338..1663
       /standard_name="NucI"
       /note="member of steroid hormone receptor superfamily;
       putative"
       /product="peroxisome proliferator activated receptor"

```
                     /codon_start=1
MMPPAR.PE1        Location/Qualifiers   (length= 1407 bp)
                  CDS          167..1573
                     /product="peroxisome proliferator activated receptor"
                     /codon_start=1
RATPPAR   Rattus rattus peroxisome proliferator activated receptor (PPAR)
          mRNA, complete cds.                        =2022 bp
RATPPAR.PPAR     Location/Qualifiers   (length= 1407 bp)
                  CDS          378..1784
                     /gene="PPAR"
                     /product="peroxisome proliferator activated receptor"
                     /codon_start=1
XELPPARA                                          Xenopus laevis
peroxisome proliferator activated receptor alpha
                  (PPARa) mRNA, complete cds.            =3250 bp
XELPPARA.PPARA    Location/Qualifiers   (length= 1425 bp)
                  CDS          108..1532
                     /gene="PPARa"
                     /product="peroxisome proliferator activated receptor
                     alpha"
                     /codon_start=1
XELPPARB   Xenopus laevis peroxisome proliferator activated receptor beta
                  (PPARb) mRNA, complete cds.            =1537 bp
XELPPARB.PPARB    Location/Qualifiers   (length= 1191 bp)
                  CDS          75..1265
                     /gene="PPARb"
                     /product="peroxisome proliferator activated receptor beta"
                     /codon_start=1
XELPPARG   Xenopus laevis peroxisome proliferator activated receptor gamma
                  (PPARg) mRNA, complete cds.            =2046 bp
XELPPARG.PPARG    Location/Qualifiers   (length= 1434 bp)
                  CDS          405..1838
                     /gene="PPARg"
                     /product="peroxisome proliferator activated receptor
                     gamma"
                     /codon_start=1
```

Command? (H for command list)
in
List name, sequence, or accession #, or H(elp)? âdefault=LIST1ê
humppara

```
LOCUS    HUMPPARA    3301 bp ss-mRNA      PRI    08-DEC-1992
DEFINITION Human peroxisome proliferator activated receptor mRNA, complete
          cds.
ACCESSION L07592
KEYWORDS  nuclear receptor; peroxisome proliferator activated receptor;
          steroid hormone receptor; steroid receptor superfamily;
```

xenobiotic responsive element.
SOURCE   Homo sapiens (library: Lambda gt11) cDNA to mRNA.
  ORGANISM Homo sapiens
            Eukaryota; Animalia; Chordata; Vertebrata; Mammalia; Theria;
            Eutheria; Primates; Haplorhini; Catarrhini; Hominidae.
REFERENCE 1 (bases 1 to 3301)
  AUTHORS Schmidt,A., Endo,N., Rutledge,S., Vogel,R., Shinar,D. and
            Rodan,G.A.
  TITLE   Identification of a new member of the steroid hormone receptor
            superfamily that is activated by a peroxisome proliferator and fatty acids
  JOURNAL Mol. Endocrinol. 6, 1634-1641 (1992)
  STANDARD full automatic
FEATURES   Location/Qualifiers
.NUCI CDS   338..1663
            /standard_name="NucI"
            /note="member of steroid hormone receptor superfamily; putative"
            /product="peroxisome proliferator activated receptor"
            /codon_start=1
                    /translation="MEQPQEEAPEVREEEEKEEVAFAEGAPELNGGPQHALPSSSYTD
            LSRSSSPPSLLDQLQMGCDGASCGSINMECRVCGDKASGFHYGVHACEGCKGFFRRTI
            RMKLEYEKCERSCKIQKKNRNKCQYCRFQKCLALGMSHNAIRFGRMPEAEKRKLVAGL
            TANEGSQYNPQVADLKAFSKHIYNAYLKNFNMTKKKARSILTGKASHTAPFVIHDIET
            LWQAEKGLVWKQLVNGLPPYKEISVHVFYRCQCTTVETVRELTEFAKSIPSFSSLFLN
            DQVTLLKYGVHEAIFAMLASIVNKDGLLVANGSGFVTREFLRSLRKPFSDIIEPKFEF
            AVKFNALELDDSDLALFIAAIILCGDRPGLMNVPRVEAIQDTILRALEFHLQANHPDA
            QYLFPKLLQKMADLRQLVTEHAQMMQRIKKTETETSLHPLLQEIYKDMY"
BASE COUNT   705 a   1014 c   879 g   696 t   7 others

Graphical representation of these sequences (menu 1-3)

This program gives some characteristics and a graphical representation of the
sequences found in EMBL or GenBank. Application to PPAR sequences found in
search A_2:

Graphical representation of characteristics of the sequence of positions		
MUSPPAR.PE1 (GenBank)	1 to 1407	1407 p

```
          1   176  352  528  704  880 1056 1232 1408
Charact.  +------+------+------+------+------+------+------+------+ from   to
mRNA      +------------------------------------------------------- 1      2081
Pept          +----------------------------------------------------- 167    1573
```

7.3.2.3 Search A₃: NBRF-PIR - Peroxisome Proliferator Activated Receptor
(PPAR)

By asking NBRF-PIR, a protein data bank, find information on the mouse PPAR
(Menu 1-1).

ACNUC Data Base Content
NBRF-PIR Release 33 (June 30th, 1992)
12,410,782 amino acids; 10,071 (+32,143 in preparation) sequences; 26,710 refs.
Research Program : M. Gouy, Laboratory of Biometry, University of Lyon 1

ENTRY	S11659 #Type Protein
TITLE	***Peroxisome proliferator-activated receptor - Mouse**
DATE	15-Jun-1992 #Sequence 15-Jun-1992 #Text 15-Jun-1992
PLACEMENT	0.0 0.0 0.0 0.0 0.0
COMMENT	*This entry is not verified.
SOURCE	Mus musculus #Common-name house mouse
REFERENCE	
#Authors	Issemann I., Green S.
#Journal	Nature (1990) 347:645-650
#Title	Activation of a member of the steroid hormone receptor superfamily by peroxisome proliferators.
#Reference-number	S11659
#Accession	S11659
SUMMARY	#Molecular-weight 52432 #Length 468 #Checksum 4196
ENTRY	S11659 #Type Protein
TITLE	***Peroxisome proliferator-activated receptor - Mouse**

```
SEQUENCE        5        10       15       20       25       30
    1 M V D T E S P I C P L S P L E A D D L E S P L S E E F L Q E
   31 M G N I Q E I S Q S I G E E S S G S F G F A D Y Q Y L G S C
   61 P G S E G S V I T D T L S P R S S P S S V S C P V I P A S T
   91 D E S P G S A L N I E C R I C G D K A S G Y H Y G V H A C E
  121 G C K G F F R R T I R L K L V Y D K C D R S C K I Q K K N R
  151 N K C Q Y C R F H K C L S V G M S H N A I R F G R M P R S E
  181 K A K L K A E I L T C E H D L K D S E T A D L K S L G K R I
  211 H E A Y L K N F N M N K V K A R V I L A G K T S N N P P F V
  241 I H D M E T L C M A E K T L V A K M V A N G V E D K E A E V
  271 R F F H C C Q C M S V E T V T E L T E F A K A I P G F A N L
  301 D L N D Q V T L L K Y G V Y E A I F T M L S S L M N K D G M
  331 L I A Y G N G F I T R E F L K N L R K P F C D I M E P K F D
  361 F A M K F N A L E L D D S D I S L F V A A I I C C G D R P G
  391 L L N I G Y I E K L Q E G I V H V L K L H L Q S N H P D D T
  421 F L F P K L L Q K M V D L R Q L V T E H A Q L V Q V I K K T
  451 E S D A A L H P L L Q E I Y R D M Y
```

7.3.2.4 Search A4: PROSITE DATA File - Peroxisomal Targeting Sequences (PTS)

By using PROSITE DATA File, a protein pattern bank (803 motifs), try to find if there are any peroxisomal targeting sequences (PTS) (menu 1-9). Use "microbodies" as the keyword.

ID   MICROBODIES_CTER; PATTERN.
AC   PS00342;
DT   NOV-1990 (CREATED); DEC-1991 (DATA UPDATE); DEC-1992 (INFO UPDATE).
DE   Microbodies C-terminal targeting signal
PA   âSAGCê-âRKHê-âLIVMAFê>.
NR   /RELEASE=24,28154;
CC   /TAXO-RANGE=??E??; /MAX-REPEAT=1;
CC   /SKIP-FLAG=TRUE;

Microbodies C-terminal targeting signal

Microbodies are a class of small organelles bound by a single membrane to which belong peroxisomes, glyoxysomes, and glycosomes. Microbodies proteins are synthesized on free polysomes and are imported into the organelle post translationally. Unlike the import of proteins into mitochondria, chloroplast or the ER/secretion pathway, import into the microbodies does not generally requires the removal of a presequence â1ê. It has been experimentally shown â2,3,4ê that, in some peroxosimal proteins, the targeting signal (PTS) resides in the last three amino acids of the C-terminus. This consensus sequence is known as "S-K-L" (Ser-Lys-Leu), but some variations are allowed in all three positions. As the peroxisomal targeting signal seems also to be recognized by other microbodies, it is now â1ê known as the C-terminal microbodies targeting signal (CMTS).

It must be noted that not all microbodies proteins contain a CMTS; some seem to contain an internal CMTS-like sequence, but it is not yet known if it is active as such; finally a few proteins are synthesized with a N-terminal presequence which is cleaved off during import.

The microbodies proteins that are known or thought to contain a CMTS are listed below.
- Mammalian D-amino acid oxidase.
- Mammalian acyl-coenzyme A oxidase (but not the fungal enzymes).
- Mammalian trifunctional fatty acid beta oxidation pathway enzyme.
- Mammalian, insect, plants, and Aspergillus uricase.
- Mammalian sterol carrier protein-2 high molecular form (SCP-X).
- Mammalian long chain alpha-hydroxy acid oxidase.
- Firefly luciferase.

- Plants glycolate oxidase.
- Plants glyoxisomal isocitrate lyase.
- Plants glyoxisomal malate synthase.
- Trypanosoma glycosomal glucose-6-phosphate isomerase.
- Trypanosoma glycosomal glyceraldehyde 3-phosphate dehydrogenase.
- Yeast (H. polymorpha and Pichia pastoris) alcohol oxidase (AOX).
- Yeast (H. polymorpha) dihydroxy-acetone synthase (DHAS).
- Yeast (S. cerevisiae) catalase A.
- Yeast (S. cerevisiae) citrate synthase.
- Yeast (C. boidinii) peroxisomal protein PMP20.
- Yeast (C. tropicalis) hydratase-dehydrogenase-epimerase (HDE) from fatty
  acid beta oxidation pathway.
- Yeast (C. tropicalis) isocitrate lyase.

-Consensus pattern: âSAGCê-âRKHê-âLIVMAFê>
-Last update: December 1992 / Text revised.

â 1ê De Hoop M.J., Ab G.
    Biochem. J. 286:657-669(1992).
â 2ê Gould S.J., Keller G.-A., Subramani S.
    J. Cell Biol. 107:897-905(1988).
â 3ê Gould S.J., Keller G.-A., Hosken N., Wilkinson J., Subramani S.
    J. Cell Biol. 108:1657-1664(1989).
â 4ê Gould S.J., Keller G.-A., Schneider M., Howell S.H., Garrard L.J.,
    Goodman J.M., Distel B, Tabak H., Subramani S.
    EMBO J. 9:85-90(1990).

Total number of found motifs :   1

7.3.3 Part B - Translation of Nucleic Sequences into Protein Sequences and Saving

As an example, translate the four nucleic sequences of the PPAR from human, rat, mouse and Xenopus into protein sequences and save them for further use.
Example for rat PPAR :

Menu 7-3 : Nucleic sequences into protein sequences

Sequence : RATPPAR.PPAR   Position   1 a 1407

```
 M  V  D  T  E  S  P  I  C  P  L  S  P  L  E  A  D  D  L  E  S  P  L  S  E
ATGGTGGACACAGAGAGCCCCATCTGTCCTCTCTCTCCCCACTTGAAGCAGATGACCTGGAAAGTCCCTTATCTGAA
 1        11        21        31        41        51        61        71
```

```
  E   F   L   Q   E   M   G   N   I   Q   E   I   S   Q   S   L   G   E   E   S   S   G   S   F   S
GAATTCTTACAAGAGATGGGAAACATTCAAGAGATTTCTCAGTCCCTCGGAGAGGAGAGTTCCGGAAGCTTTAGT
76        86        96        106       116       126       136       146
  F   A   D   Y   Q   Y   L   G   S   C   P   G   S   E   G   S   V   I   T   D   T   L   S   P   A
TTTGCGGACTACCAGTACTTAGGGAGCTGTCCAGGCTCGGAGGGCTCTGTCATCACAGACACCCTCTCTCCAGCT
151       161       171       181       191       201       211       221
  S   S   P   S   S   V   S   C   P   A   V   P   T   S   T   D   E   S   P   G   N   A   L   N   I
TCCAGCCCCTCCTCAGTCAGCTGCCCTGCTGTCCCCACCAGTACAGATGAGTCCCTGGCAATGCACTGAACATC
226       236       246       256       266       276       286       296
  E   C   R   I   C   G   D   K   A   S   G   Y   H   Y   G   V   H   A   C   E   G   C   K   G   F
GAGTGTCGAATATGTGGGGACAAGGCCTCAGGATACCACTATGGAGTCCACGCATGTGAAGGCTGCAAGGGCTTC
301       311       321       331       341       351       361       371
  F   R   R   T   I   R   L   K   L   A   Y   D   K   C   D   R   S   C   K   I   Q   K   K   N   R
TTTCGGCGAACTATTCGGCTAAAGCTGGCGTACGACAAGTGTGATCGAAGCTGCAAGATTCAGAAAAAGAACCGG
376       386       396       406       416       426       436       446
  N   K   C   Q   Y   C   R   F   H   K   C   L   S   V   G   M   S   H   N   A   I   R   F   G   R
AACAAATGCCAGTACTGCCGTTTCCACAAGTGCCTGTCCGTCGGGATGTCACACAATGCAATCCGTTTTGGAAGA
451       461       471       481       491       501       511       521
  M   P   R   S   E   K   A   K   L   K   A   E   I   L   T   C   E   H   D   L   K   D   S   E   T
ATGCCAAGATCTGAGAAAGCAAAACTGAAGGCAGAAATCCTTACCTGTGAACACGATCTGAAAGATTCGGAAACT
526       536       546       556       566       576       586       596
  A   D   L   K   S   L   A   K   R   I   H   E   A   Y   L   K   N   F   N   M   N   K   V   K   A
GCAGACCTCAAATCTCTGGCCAAGAGAATCCACGAAGCCTACCTGAAGAACTTCAACATGAACAAGGTCAAGGCC
601       611       621       631       641       651       661       671
  R   V   I   L   A   G   K   T   S   N   N   P   P   F   V   I   H   D   M   E   T   L   C   M   A
CGGGTCATACTCGCAGGAAAGACTAGCAACAATCCGCCTTTTGTCATACATGACATGGAGACCTTGTGCATGGCT
676       686       696       706       716       726       736       746
  E   K   T   L   V   A   K   M   V   A   N   G   V   E   N   K   E   A   E   V   R   F   F   H   C
GAGAAGACGCTTGTGGCCAAGATGGTAGCCAACGGCGTTGAAAACAAGGAGGCAGAGGTCCGATTCTTCCACTGC
751       761       771       781       791       801       811       821
  C   Q   C   M   S   V   E   T   V   T   E   L   T   E   F   A   K   A   I   P   G   F   A   N   L
TGCCAGTGCATGTCCGTGGAGACCGTCACCGAGCTCACGGAATTTGCCAAGGCTATCCCAGGCTTTGCAAACTTG
826       836       846       856       866       876       886       896
  D   L   N   D   Q   V   T   L   L   K   Y   G   V   Y   E   A   I   F   T   M   L   S   S   L   M
GACTTGAATGACCAGGTTACCTTGCTGAAGTACGGTGTGTATGAAGCCATCTTCACGATGCTGTCCTCCTTGATG
901       911       921       931       941       951       961       971
  N   K   D   G   M   L   I   A   Y   G   N   G   F   I   T   R   E   F   L   K   N   L   R   K   P
AACAAAGACGGGATGCTGATCGCGTACGGCAATGGCTTCATCACCCGAGAGTTCCTAAAGAACCTGAGGAAGCCA
976       986       996       1006      1016      1026      1036      1046
  F   C   D   I   M   E   P   K   F   D   F   A   M   K   F   N   A   L   E   L   D   D   S   D   I
TTCTGCGACATCATGGAACCCAAGTTTGACTTCGCTATGAAGTTCAATGCCCTCGAACTGGATGACAGTGACATT
1051      1061      1071      1081      1091      1101      1111      1121
```

```
   S  L  F  V  A  A  I  I  C  C  G  D  R  P  G  L  L  N  I  G  Y  I  E  K'L
TCCCTTTTTGTGGCTGCTATAATTTGCTGTGGAGATCGGCCTGGCCTTCTAAACATAGGATACATTGAGAAGTTG
1126      1136      1146      1156      1166      1176      1186      1196
   Q  E  G  I  V  H  V  L  K  L  H  L  Q  S  N'H  P  D  D  T  F  L  F  P  K
CAGGAGGGGATTGTGCACGTGCTCAAGCTCCACCTGCAGAGCAACCATCCGGATGATACCTTTCTCTTCCCAAAA
1201      1211      1221      1231      1241      1251      1261      1271
   L  L  Q  K  M  V  D  L  R  Q  L  V  T  E  H  A  Q  L  V  Q  V  I  K  K  T
CTCCTTCAAAAAATGGTGGACCTCCGGCAGCTGGTCACGGAGCATGCGCAGCTCGTGCAGGTCATCAAGAAGACC
1276      1286      1296      1306      1316      1326      1336      1346
   E  S  D  A  A  L  H  P  L  L  Q  E  I  Y  R  D  M  Y  *
GAGTCAGACGCGGCGTTGCACCCACTGTTGCAAGAGATCTACAGAGACATGTACTGA
1351      1361      1371      1381      1391      1401
```

7.3.4   Part C -   Alignments of Protein Sequences

7.3.4.1 Search C_1: Alignments of Two Sequences

The menu 5-6 allows you to align two protein sequences (Kanehisa algorithm). These alignments are possible only if a minimum number of consecutive amino acids are homologous. We will compare rat and mouse PPAR and then mouse and Xenopus PPAR or human and rat PPAR by working on previous saved sequences (from Part B).

Example for mouse and rat PPAR:

Menu 5-6 : Homologies and alignment of the two protein sequences of mouse and rat PPAR (Kanehisa)

PARAMETERS ARE:
 MAXD = -50   KTPL = 2   NW = 25   LWID = 70
 DEL = 8   BIAS = 0   MATRIX = MD

```
QUERY SEQUENCE    TRADUCMUSPPAR      FROM    1  TO   468  TOTAL   468

================================================================================

QUERY SEQUENCE    TRADUCRATPPAR      FROM    1  TO   468  TOTAL   468


        10        20        30        40        50        60        70
MVDTESPICPLSPLEADDLESPLSEEFLQEMGNIQEISQSIGEESSGSFGFADYQYLGSCPGSEGSVITD
                                       --========----=
MVDTESPICPLSPLEADDLESPLSEEFLQEMGNIQEISQSLGEESSGSFSFADYQYLGSCPGSEGSVITD
        10        20        30        40        50        60        70
```

```
          80        90       100       110       120       130       140
TLSPRSSPSSVSCPVIPASTDESPGSALNIECRICGDKASGYHYGVHACEGCKGFFRRTIRLKLVYDKCD
===  ==========  --=-========  =============================  ===  ====
TLSPASSPSSVSCPAVPTSTDESPGNALNIECRICGDKASGYHYGVHACEGCKGFFRRTIRLKLAYDKCD
          80        90       100       110       120       130       140

         150       160       170       180       190       200       210
RSCKIQKKNRNKCQYCRFHKCLSVGMSHNAIRFGRMPRSEKAKLKAEILTCEHDLKDSETADLKSLGKRI
=====================================================================
RSCKIQKKNRNKCQYCRFHKCLSVGMSHNAIRFGRMPRSEKAKLKAEILTCEHDLKDSETADLKSLAKRI
         150       160       170       180       190       200       210

         220       230       240       250       260       270       280
HEAYLKNFNMNKVKARVILAGKTSNNPPFVIHDMETLCMAEKTLVAKMVANGVEDKEAEVRFFHCCQCMS
=========================================================--===========
HEAYLKNFNMNKVKARVILAGKTSNNPPFVIHDMETLCMAEKTLVAKMVANGVENKEAEVRFFHCCQCMS
         220       230       240       250       260       270       280

         290       300       310       320       330       340       350
VETVTELTEFAKAIPGFANLDLNDQVTLLKYGVYEAIFTMLSSLMNKDGMLIAYGNGFITREFLKNLRKP
=====================================================================
VETVTELTEFAKAIPGFANLDLNDQVTLLKYGVYEAIFTMLSSLMNKDGMLIAYGNGFITREFLKNLRKP
         290       300       310       320       330       340       350

         360       370       380       390       400       410       420
FCDIMEPKFDFAMKFNALELDDSDISLFVAAIICCGDRPGLLNIGYIEKLQEGIVHVLKLHLQSNHPDDT
=====================================================================
FCDIMEPKFDFAMKFNALELDDSDISLFVAAIICCGDRPGLLNIGYIEKLQEGIVHVLKLHLQSNHPDDT
         360       370       380       390       400       410       420

         430       440       450       460
FLFPKLLQKMVDLRQLVTEHAQLVQVIKKTESDAALHPLLQEIYRDMY
===============================================
FLFPKLLQKMVDLRQLVTEHAQLVQVIKKTESDAALHPLLQEIYRDMY
         430       440       450       460


> SEQ 1:TRADUCMUSPPAR   ( 1-   468)     SEQ 2:TRADUCRATPPAR   ( 1-    468)
> HOMOLOGY SCORE= -2328   (-2328)    % MATCH=  97.9 ( 458 / 468)
```

7.3.4.2 Search C₂: *Alignments of Several Sequences*

The program 5-9 allows you to align sequences of similar size and supposed
homeologies.

Application to the four translated saved sequences of human, rat, mouse and
Xenopus PPAR (from part B):
- To start, format the sequences (menu 1-12, choice 8).
- Then, execute the alignment program (menu 5-9).

- Finally, recuperate the files created in menu 10-1 file managing.

CLUSTAL V ... Multiple Sequence Alignments

Sequence 1: TRADUC HUM PPAR   441 aa
Sequence 2: TRADUC MUS PPAR   468 aa
Sequence 3: TRADUC RAT PPAR   468 aa
Sequence 4: TRADUC XEL PPAR   474 aa

Sequences (1:2) Aligned. Score: 60.3175
Sequences (1:3) Aligned. Score: 60.0907
Sequences (1:4) Aligned. Score: 61.2245
Sequences (2:3) Aligned. Score: 97.8632
Sequences (2:4) Aligned. Score: 76.7094
Sequences (3:4) Aligned. Score: 76.9231

CLUSTAL V multiple sequence alignment

```
TRADUCHUMP    M----------------EQPQEEAP-------EVREEEEKEEVAEAEGAPELN------
TRADUCMUSP    M----VDTESPICPLSPLEADDLESPLSEEFLQEMGNIQEISQSIGEESSGSFGFADYQY
TRADUCRATP    M----VDTESPICPLSPLEADDLESPLSEEFLQEMGNIQEISQSLGEESSGSFSFADYQY
TRADUCXELP    MSTIMVDTNSFLCILTPLDEDDLESPLSGEFLQDIVDIQDITQTIGDDGSTPFGASEHQF
              *                  .   .*.*        ..  .  ..    ...  .

TRADUCHUMP    ----GGPQHALPSSSYTDLSRSSSPPSLL-DQLQMGCDGASCGSLNMECRVCGDKASGFH
TRADUCMUSP    LGSCPGSEGSV---ITDTLSPRSSPSSVSCPVIPASTDESPGSALNIECRICGDKASGYH
TRADUCRATP    LGSCPGSEGSV---ITDTLSPASSPSSVSCPAVPTSTDESPGNALNIECRICGDKASGYH
TRADUCXELP    FGNSPGSIGSVSTDLTDTLSPASSPASITFPAASGSAEDAACKSLNLECRVCSDKASGFH
                 *   ..      **   ***.*.     .   .. .**.***.*.*****.*

TRADUCHUMP    YGVHACEGCKGFFRRTIRMKLEYEKCERSCKIQKKNRNKCQYCRFQKCLALGMSHNAIRF
TRADUCMUSP    YGVHACEGCKGFFRRTIRLKLVYDKCDRSCKIQKKNRNKCQYCRFHKCLSVGMSHNAIRF
TRADUCRATP    YGVHACEGCKGFFRRTIRLKLAYDKCDRSCKIQKKNRNKCQYCRFHKCLSVGMSHNAIRF
TRADUCXELP    YGVHACEGCKGFFRRTIRLKLVYDRCERMCKIQKKNRNKCQYCRFEKCLNVGMSHNAIRF
              *****************.**  *..*.* ****************.***  .*********

TRADUCHUMP    GRMPEAEKRKLVAGLTANEGSQYNPQVADLKAFSKHIYNAYLKNFNMTKKKARSILTGKA
TRADUCMUSP    GRMPRSEKAKLKAEIILTCEHDLKDSETADLKSLGKRIHEAYLKNFNMNKVKARVILAGKT
TRADUCRATP    GRMPRSEKAKLKAEILTCEHDLKDSETADLKSLAKRIHEAYLKNFNMNKVKARVILAGKT
TRADUCXELP    GRMPRSEKAKLKAEVLMCDQDVKDSQMADLLSLARLIYDAYLKNFNMNKVKARAILTGKA
              ****  ** ** *    ...  ***  ....  * .******** * *** **.**.

TRADUCHUMP    SHTAPFVIHDIETLWQAEKGLVWKQLVNGLPPYKEISVHVFYRCQCTTVETVRELTEFAK
TRADUCMUSP    SNNPPFVIHDMETLCMAEKTLVAKMVANGVED-KEAEVRFFHCCQCMSVETVTELTEFAK
TRADUCRATP    SNNPPFVIHDMETLCMAEKTLVAKMVANGVEN-KEAEVRFFHCCQCMSVETVTELTEFAK
TRADUCXELP    SN-PPFVIHDMETLCMAEKTLVAKLVANGIQN-KEAEVRIFHCCQCTSVETVTELTEFAK
              *. .******.***  *** **  *  .**.    **  *. *   *** .**** *******

TRADUCHUMP    SIPSFSSLFLNDQVTLLKYGVHEAIFAMLASIVNKDGLLVANGSGFVTREFLRSLRKPFS
TRADUCMUSP    AIPGFANLDLNDQVTLLKYGVYEAIFTMLSSLMNKDGMLIAYGNGFITREFLKNLRKPFC
TRADUCRATP    AIPGFANLDLNDQVTLLKYGVYEAIFTMLSSLMNKDGMLIAYGNGFITREFLKNLRKPFC
TRADUCXELP    SIPGFTELDLNDQVTLLKYGVYEAMFAMLASVMNKDGMLVAYGNGFITREFLKSLRKPIG
              .**.*. * ************ **.*.**.. .****.*.* *.**.***** .****.
```

```
TRADUCHUMP   DIIEPKFEFAVKFNALELDDSDLALFIAAIILCGDRPGLMNVPRVEAIQDTILRALEFHL
TRADUCMUSP   DIMEPKFDFAMKFNALELDDSDISLFVAAIICCGDRPGLLNIGYIEKLQEGIVHVLKLHL
TRADUCRATP   DIMEPKFDFAMKFNALELDDSDISLFVAAIICCGDRPGLLNIGYIEKLQEGIVHVLKLHL
TRADUCXELP   DMMEPKFEFAMKFNALELDDSDLSLFVAALICCGDRPGLVNIPSIEKMQESIVHVLKLHL
             *..****.**.***********..**.**.*.*******.*..*..*.*..*..*..**

TRADUCHUMP   QANHPDAQYLFPKLLQKMADLRQLVTEHAQMMQRIKKTETETSLHPLLQEIYKDMY
TRADUCMUSP   QSNHPDDTFLFPKLLQKMVDLRQLVTEHAQLVQVIKKTESDAALHPLLQEIYRDMY
TRADUCRATP   QSNHPDDTFLFPKLLQKMVDLRQLVTEHAQLVQVIKKTESDAALHPLLQEIYRDMY
TRADUCXELP   QSNHPDDSFLFPKLLQKMADLRQLVTEHAQLVQTIKKTETDAALHPLLQEIYRDMY
             *.****.*********.***********..*.*****....*********.***
```

7.3.5 Part D - Primary Structure of Nucleic Sequences

Starting from a nucleic sequence, it is possible to obtain information on:
- frequencies of bases and codons
- consensus sequences
- repeating sequences

7.3.5.1 Search D_1: Base and Codon Frequencies

Program 3-1 permits the calculation of base and codon frequencies in a given sequence. Application to acyl CoA oxidase mRNA from rat (locus : RNACOA1 in EMBL):

Menu 3-1- option 1 : Base frequency
Application to acyl CoA oxidase mRNA from rat

RNACOA1 POSITIONS   1 a 3741
Base frequency (number and percentage)

POSN	1		2 2		3		TOTAL	
A	269	21.6%	345	27.7%	367	29.4%	981	26.2%
C	332	26.6%	304	24.4%	272	21.8%	908	24.3%
G	338	27.1%	322	25.8%	234	18.8%	894	23.9%
T	308	24.7%	276	22.1%	374	30.0%	958	25.6%

Menu 3-1- option 6 :
Potential amino acid frequency in number and in %/total
Application to acyl CoA oxidase mRNA from rat

RNACOA1 POSITIONS   3 to 3741

AMINO ACID		PHASE 1		PHASE 2		PHASE 3	
		NBR	%/TOTAL	NBR	%/TOTAL	NBR	%/TOTAL
ALA	A	76	6.1%	69	5.5%	65	5.2%
ARG	R	79	6.3%	63	5.1%	86	6.9%
ASN	N	38	3.0%	48	3.9%	36	2.9%
ASP	D	42	3.4%	37	3.0%	20	1.6%
CYS	C	50	4.0%	23	1.8%	57	4.6%
GLN	Q	57	4.6%	51	4.1%	42	3.4%
GLU	E	64	5.1%	62	5.0%	37	3.0%
GLY	G	93	7.5%	71	5.7%	62	5.0%
HIS	H	53	4.3%	46	3.7%	34	2.7%
ILE	I	34	2.7%	55	4.4%	45	3.6%
LEU	L	125	10.0%	151	12.1%	129	10.4%
LYS	K	43	3.4%	61	4.9%	50	4.0%
MET	M	12	1.0%	30	2.4%	35	2.8%
PHE	F	42	3.4%	55	4.4%	50	4.0%
PRO	P	94	7.5%	62	5.0%	88	7.1%
SER	S	110	8.8%	113	9.1%	126	10.1%
THR	'T	62	5.0%	73	5.9%	96	7.7%
TRP	W	29	2.3%	19	1.5%	43	3.5%
TYR	Y	25	2.0%	45	3.6%	17	1.4%
VAL	V	63	5.1%	83	6.7%	49	3.9%
TAA	*	15	1.2%	13	1.0%	23	1.8%
TAG	*	8	0.6%	4	0.3%	9	0.7%
TGA	*	33	2.6%	12	1.0%	47	3.8%
TOTAL CODONS		1247		1246		1246	

7.3.5.2 Search D2: Consensus Sequences

In menu 3-15, search for the consensus sequence TGACCT N TGTCCT in the 5'
upstream region of rat acyl CoA oxidase (locus : RNACOA2 in EMBL).

Menu 3- 15: Sequence primary structure
Search for a consensus nucleic sequence (TFD, GNOMIC)

Consensus TGACCTNTGTCCT
Probability correct motif = 0.5960E-07   Predicted number =   0.00

1. SCORE 13   704 - 716   TGACCTTTGTCCT

704   714

TGACCTTTGTCCT

* * * * * * * * * * * * *

7.3.5.3 Search D₃: Repeating Sequences

Menu 4-1 : Search for a direct repeating sequence in the 5' upstream region of the rat acyl CoA oxidase gene (locus : RNACOA2 in EMBL).

Menu. 4-1: Search for a direct nucleic sequence repetition

Mnemonique of the sequence : rnacoa2
MINIMAL LENGTH OF THE REPEATING SEQUENCE ?
15
NUMBER OF REPETITIONS = 1
SCORE   17
POSIT. 1   373
POSIT. 2   482

LG = 17   373-AAAGAAATGTAGTTCTT
482-AAAGAAATGTAGTTCTT

7.3.6 Part E - Protein Data Analysis

7.3.6.1 Search E₁: Information, Composition and Sequence of a Protein

Program 8-1, applied to the mouse PPAR (NBRF locus : S11659), will list the comments, edit the amino acid composition and represent the protein sequence.

Menu 8-1 : Information, composition and protein list

ENTRY	S11659 #Type Protein
TITLE	*Peroxisome proliferator-activated receptor - Mouse
DATE	15-Jun-1992 #Sequence 15-Jun-1992 #Text 15-Jun-1992
PLACEMENT	0.0 0.0 0.0 0.0 0.0
COMMENT	*This entry is not verified.
SOURCE	Mus musculus #Common-name house mouse
REFERENCE	
#Authors	Issemann I., Green S.
#Journal	Nature (1990) 347:645-650
#Title	Activation of a member of the steroid hormone receptor

110

superfamily by peroxisome proliferators.
#Reference-number S11659
#Accession   S11659
SUMMARY   #Molecular-weight 52432 #Length 468 #Checksum 4196

ALA A	29 (6.2)	CYS C	20 (4.3)	ASP D	28 (6.0)	GLU E	35 (7.5)
PHE F	23 (4.9)	GLY G	27 (5.8)	HIS H	13 (2.8)	ILE I	28 (6.0)
LYS K	35 (7.5)	LEU L	48 (10.3)	MET M	16 (3.4)	ASN N	19 (4.1)
PRO P	20 (4.3)	GLN Q	15 (3.2)	ARG R	19 (4.1)	SER S	36 (7.7)
THR T	19 (4.1)	VAL V	25 (5.3)	TRP W	0 (0.0)	TYR Y	13 (2.8)
ASX B	0 (0.0)	GLX Z	0 (0.0)	*** *	0 (0.0)		

RESIDUES   468
OTAL MW =   52398.17
POLARITY INDEX =   46.79
ELATIVE BASICITY (H+K+R/D+E) = 1.06

```
              5        10       15       20       25       30
        M V D T E S P I C P L S P L E A D D L E S P L S E E F L Q E
             35       40       45       50       55       60
        M G N I Q E I S Q S I G E E S S G S F G F A D Y Q Y L G S C
             65       70       75       80       85       90
        P G S E G S V I T D T L S P R S S P S S V S C P V I P A S T
             95      100      105      110      115      120
        D E S P G S A L N I E C R I C G D K A S G Y H Y G V H A C E
            125      130      135      140      145      150
        G C K G F F R R T I R L K L V Y D K C D R S C K I Q K K N R
            155      160      165      170      175      180
        N K C Q Y C R F H K C L S V G M S H N A I R F G R M P R S E
            185      190      195      200      205      210
        K A K L K A E I L T C E H D L K D S E T A D L K S L G K R I
            215      220      225      230      235      240
        H E A Y L K N F N M N K V K A R V I L A G K T S N N P P F V
            245      250      255      260      265      270
        I H D M E T L C M A E K T L V A K M V A N G V E D K E A E V
            275      280      285      290      295      300
        R F F H C C Q C M S V E T V T E L T E F A K A I P G F A N L
            305      310      315      320      325      330
        D L N D Q V T L L K Y G V Y E A I F T M L S S L M N K D G M
            335      340      345      350      355      360
        L I A Y G N G F I T R E F L K N L R K P F C D I M E P K F D
            365      370      375      380      385      390
        F A M K F N A L E L D D S D I S L F V A A I I C C G D R P G
            395      400      405      410      415      420
```

111

```
L L N I G Y I E K L Q E G I V H V L K L H L Q S N H P D D T
    425         430         435         440         445         450
F L F P K L L Q K M V D L R Q L V T E H A Q L V Q V I K K T
    455         460         465
.E S D A A L H P L L Q E I Y R D M Y
```

7.3.6.2 Search E_2: Molecular Mass of a Protein

Calculate the amino acid composition (number, percentage and mass) of the mouse PPAR (NBRF locus : S11659). This will give polarity, theoretical isoelectric point and theoretical absorbance at 260 and 280nm (menu 8-2).

Menu 8-2: Protein molecular mass

			NUMBER	% NOMB	WEIGHT	% WEIGHT
1	Phe	F	23	4.91	3382.57	6.46
2	Leu	L	48	10.26	5428.03	10.36
3	Ile	I	28	5.98	3166.35	6.04
4	Met	M	16	3.42	2096.65	4.00
5	Val	V	25	5.34	2476.71	4.73
6	Ser	S	36	7.69	3133.15	5.98
7	Pro	P	20	4.27	1941.06	3.70
8	Thr	T	19	4.06	1919.91	3.66
9	Ala	A	29	6.20	2060.08	3.93
10	Tyr	Y	13	2.78	2119.82	4.05
12	His	H	13	2.78	1781.77	3.40
13	Gln	Q	15	3.21	1920.88	3.67
14	Asn	N	19	4.06	2166.82	4.14
15	Lys	K	35	7.48	4483.32	8.56
16	Asp	D	28	5.98	3220.75	6.15
17	Glu	E	35	7.48	4516.49	8.62
18	Cys	C	20	4.27	2060.18	3.93
19	Trp	W	0	0.00	0.00	0.00
20	Arg	R	19	4.06	2965.92	5.66
21	Gly	G	27	5.77	1539.58	2.94

```
RESIDUES                    =      468
Molecular mass (monoisotopic) =    52398.1680
Molecular mass (moyenneà    =      52432.4805
POLARITY INDEX (%) =       46.79
ISOELECTRIC POINT  =        6.51
Ab surface 260 (1mg/ml) =  0.209  Ab sur 280 (1mg/ml) =  0.295
```

7.3.6.3 Search E₃: Secondary Structure of a Protein

Search for the secondary structure of a protein sequence according to Chou-Fasman algorithm (Menu 4-10) will be applied to the mouse PPAR (NBRF locus : S 11659) :

Menu 10-4 : Protein secondary structures (Chou-Fasman)

This code identifies potential protein secondary structural domains using the Chou-Fasman pseudo probabilities for both DNA and protein sequence information. For nucleic acid sequences, it will also calculate codon usage, codon boundary, codon mid base, and nearest neighbor statistics.

This code was developed by W.W. Ralph and T.F. Smith for the MBCRR, Harvard, 1985.

Chou-Fasman parameter levels to be used:
- Alpha former definition value 1.12
- Alpha threshold, 4 residue ave. 1.08
- Alpha cutoff 1.00
- Beta threshold, 3 residue ave. 1.25
- Beta cutoff 0.97
- Turn min. level 0.50

MNEMONIQUE S11659   NUMBER OF RESIDUES   468

```
MVDTESPICPLSPLEADDLESPLSEEFLQEMGNIQEISQSIGEESSGSFGFADYQYLGSC
        10        20        30        40        50        60
      aaAAAAAAAAAAAAAAAAAAAAaa       aaAAAAaa


BBB                                              BBBB
     TTTTTTTTTT    TTTT                    TTTT        T


PGSEGSVITDTLSPRSSPSSVSCPVIPASTDESPGSALNIECRICGDKASGYHYGVHACE
        70        80        90       100       110       120
                        aaAAAAAaa aaAAAAaa   aaAAAA
    BBBBBBB          BBBB                   BBBBBBo
TTT          TTTTTTTT TTTT     TTTT       TTTTTTTT      TT


GCKGFFRRTIRLKLVYDKCDRSCKIQKKNRNKCQYCRFHKCLSVGMSHNAIRFGRMPRSE
        130       140       150       160       170       180
aa      aaaaAaaa     aaAAAAaa                 aaAAaAaaaaAAA
   BBBBBBBBoBBBB                BBBBBBB BBBBB    boBoB
TTTTT              TTTT     TTTT                     TTTT
```

```
KAKLKAEILTCEHDLKDSETADLKSLGKRIHEAYLKNFNMNKVKARVILAGKTSNNPPFV
        190       200       210       220       230       240
AAAAAAAaaaaAAAAAAAAAAAAAaa aaAAAAaa   aaAAAAAAaaaaAaa          a
        BBB                               BBBB           BB
                              TTTT             TTTTTTTT


IHDMETLCMAEKTLVAKMVANGVEDKEAEVRFFHCCQCMSVETVTELTEFAKAIPGFANL
        250       260       270       280       290       300
aAAAAAAAAAAAAaaAAAaaaaAAAAAAAAaaa   aaAAAAaaaAAAAAAAaaaaAAAA
Bb          bBBb          bBBBBBBBBbb bBBB
                                              TTTT


DLNDQVTLLKYGVYEAIFTMLSSLMNKDGMLIAYGNGFITREFLKNLRKPFCDIMEPKFD
        310       320       330       340       350       360
aa   aaaaaaaaaaAAaaaaAAAAAAAAAAAaAaaa  aaaAAAAAAaa  aaAAAAAAAA
     BBBBBBBBBB BBBBb      bBBbBBB BBBB
                     TTTT    TTTT          TTTT
FAMKFNALELDDSDISLFVAAIICCGDRPGLLNIGYIEKLQEGIVHVLKLHLQSNHPDDT
        370       380       390       400       410       420
AAAAAAAAAAaa  aaAaaAaaa       aaaaAAAAAAaaaaaaAaaa          a
             BbBBbBBBBB      BBBBBBB      BBBBBBbBBB          B
     TTTT              TTTT TTTT              TTTT


FLFPKLLQKMVDLRQLVTEHAQLVQVIKKTESDAALHPLLQEIYRDMY
        430       440       450       460
aaaAAAAAAAAAAaaaaaAAAAaaaaaAAAAAAAAAAAAAAAAAaa
BBB          BBBBB  bBBBBBb              bBBB
   TTTT                              TTTT
```

7.3.6.4 Search E4: Minimum Degenerescence Probes

From the rat peroxisomal acyl-CoA oxidase I A chain (locus OXRTA1) in NBRF-PIR, find one or several groups of minimum degenerescence probes.

Menu 7-5: Probe searching from a protein sequence

Parameters for the study of the protein sequence (locus OXRTA1):

Beginning position No ?   (1 = Return   Retour au menu = -1)
End position   No ?   (661 = Return)
Maximal degenerescence of the probe ?
15
Minimal length of the probe ? (14 bases = Return)
17

Three probe groups have been selected:
How many groups do you want to edit ?   (All = Return   None = -1)

```
N.  1      probes of 17 bases      Degen.: 12      %GC(mn-mx)   : 29- 47
           ═══════════════════                     T.hyb(mn-mx) : 44- 50
           83    84    85    86    87    88
PROT       GLU   GLU   ILE   MET   TRP   PHE
Probe 5'   GAA   GAA   ATT   ATG   TGG   TT    3'
            G     G     C
                        A
```

```
N.  2      probes of 17 bases      Degen.: 12      %GC(mn-mx)   : 23- 41
           ═══════════════════                     T.hyb(mn-mx) : 42- 48
           84    85    86    87    88    89
PROT       GLU   ILE   MET   TRP   PHE   LYS
Probe 5'   GAA   ATT   ATG   TGG   TTT   AA    3'
            G     C                 C
                  A
```

```
N.  3      probes of 17 bases      Degen.: 12      %GC(mn-mx)   : 17- 35
           ═══════════════════                     T.hyb(mn-mx) : 40- 46
           85    86    87    88    89    90
PROT       ILE   MET   TRP   PHE   LYS   LYS
Probe 5'   ATT   ATG   TGG   TTT   AAA   AA    3'
            C                 C     G
            A
```

Probe edition :
N 1 : 12 possible probes of 17 bases each
```
GAAGAAATTATGTGGTT
GAAGAAATCATGTGGTT
GAAGAAATAATGTGGTT
GAAGAGATTATGTGGTT
GAAGAGATCATGTGGTT
GAAGAGATAATGTGGTT
GAGGAAATTATGTGGTT
GAGGAAATCATGTGGTT
GAGGAAATAATGTGGTT
GAGGAGATTATGTGGTT
GAGGAGATCATGTGGTT
GAGGAGATAATGTGGTT
```

N 2 : 12 possible probes of 17 bases each
```
ATTATGTGGTTTAAAAA
ATTATGTGGTTTAAGAA
ATTATGTGGTTCAAAAA
ATTATGTGGTTCAAGAA
ATCATGTGGTTTAAAAA
ATCATGTGGTTTAAGAA
ATCATGTGGTTCAAAAA
ATCATGTGGTTCAAGAA
ATAATGTGGTTTAAAAA
ATAATGTGGTTTAAGAA
ATAATGTGGTTCAAAAA
ATAATGTGGTTCAAGAA
```

N 3 : 12 possible probes of 17 bases each

```
ATTATGTGGTTTAAAAA
ATTATGTGGTTTAAGAA
ATTATGTGGTTCAAAAA
ATTATGTGGTTCAAGAA
ATCATGTGGTTTAAAAA
ATCATGTGGTTTAAGAA
ATCATGTGGTTCAAAAA
ATCATGTGGTTCAAGAA
ATAATGTGGTTTAAAAA
ATAATGTGGTTTAAGAA
ATAATGTGGTTCAAAAA
ATAATGTGGTTCAAGAA
```

7.3.7 Some Other Working Possibilities

- Search for probability of coding region in a nucleic sequence
- Search for TATAAT signal (TATA box), promotor, signal in 3' ends of intron (AG) or 5' ends of intron (GT)
- Search for restriction maps
- Search for evidence of mutations ---> diseases
- Search for prediction of protein antigenicity

Reference

1. Dessen P, Fondrat C, Valencien C, Mugnier C (1990) Bisance: a french service for access to biomolecular sequence database. Cabios 6:355-356

Useful Addresses

GenBANK
The Computer and Information Sciences Division
BBN Laboratories Inc.
10 Moulton Street
Cambridge, Massachusetts 02238 USA
Telephone: (617) 497 2742

EMBL Data Library
European Molecular Biology Laboratory
Meyerhofstr. 1
D- 69117 Heidelberg
Telephone: 6221-387409

Protein Identification Resource
National Biomedical Research Foundation
Georgetown University Medical Center
3900 Reservoir Road, N.W.
Washington, D.C. 20007, USA
Telephone: (202) 625 2121

SWISSPROT
A. Bairoch, Departement de Biochimie Médicale
C.M.U., 1 rue Michel Servet
121 Genève 4, SWITZERLAND

C.I.T.I. 2
Centre Inter-universitaire de Traitement de l'Information
Université René Descartes
45 rue des Saints Pères
75270 Paris Cedex 06, FRANCE
Telephone: (1) 42 96 24 89

III Toxicology and Pharmacology

8 Cytochrome P450 Isoenzyme Induction Pattern in Rats Treated with Peroxisome Proliferators and Classical Inducers

M.C. CORNU, S. GREGOIRE, J.C. LHUGUENOT, M.H. SIESS, H. THOMAS, M. VILLERMAIN and F. WAECHTER

Summary

8.1 General Introduction and Aims

8.2 Quantitative Determination of Total Cytochrome P-450 in Microsomal Fractions

    8.2.1 Rat treatment

    8.2.2 Preparation of Liver Sub-Cellular Fractions

    8.2.3 Protein Assay

    8.2.4 Quantitative Determination of Total Cytochrome P-450 in Hepatic Microsomes

8.3 Induced Enzymatic Activity Assays

    8.3.1 Ethoxyresorufin O-Deethylase (EROD) and Pentoxyresorufin O-Depentylase (PROD)

    8.3.2 Peroxisomal Acyl-CoA Oxydase

    8.3.3 Lauric Acid Hydroxylase

8.4 Immunoblotting Analysis of Cytochrome P-450 Expression in Rat Liver

    8.41 Equipement, Chemicals and Solutions

    8.4.2 Experimental Procedure

8.1 General Introduction and Aims

Cytochromes P-450 are a superfamily of enzymes catalyzing the oxidation of numerous endogenous compounds (fatty acids, steroids, eicosanoids) and exogenous chemicals (drugs, pesticides, environmental pollutants). Based on DNA sequence homology, this superfamily can be further divided into enzyme families from which four families, CYP1, CYP2, CYP3 and CYP4, are involved in foreign compound metabolism and toxicity [6, 13, 20].

The CYP1A subfamily consists of only two isoenzymes which are inducible by and metabolize polycyclic aromatic hydrocarbons [15]. Typical activities of CYP1A1 are those of arylhydrocarbon hydroxylase (AHH) or ethoxyresorufin O-deethylase [3, 15, 20].

The isoenzymes of the CYP2B subfamily are characterized by their inducibility by barbiturates. Typical enzymatic activities comprise those of aminopyrene demethylase and pentoxyresorufin O-depentylase [3, 15, 20].

The CYP3A subfamily consists of at least two isoenzymes in rat liver, from which CYP3A1 is inducibly expressed in both sexes, whereas CYP3A2 is constitutively and inducibly expressed in male rats only. While CYP3A1 is induced by steroids, macrolid antibiotics and phenobarbitone-type inducers of foreign compound metabolizing enzymes, CYP3A2 induction is confined to phenobarbitone and phenobarbitone-type inducers. Diagnostic catalytic activities comprise erythromycin demethylation and testosterone 2β-, 6β- and 15β-hydroxylation [15, 20, 24].

Peroxisome proliferator inducible lauric acid 11- and 12-hydroxylase activities have been shown to reside predominantly on distinct isoenzymes of the CYP4A subfamily [1]. In rat liver, at least three members of this gene subfamily have been characterized by cDNA (CYP4A1 and CYP4A3) and/or genomic nucleotide sequencing (CYP4A1 and CYP4A2). Cytochrome CYP4A1 shares about 65% sequence similarity with CYP4A2 and 4A3, while the coding DNA sequences are 97% similar between CYP4A2 and 4A3 [1, 9, 10]. Upon SDS-gelelectrophoresis, CYP4A1 and CYP4A2 comigrate with an apparent molecular weight of 51.5 kDa, while CYP4A3 displays a slightly higher apparent molecular weight of 52 kDa [1]. On the protein and mRNA level, CYP4A2 has been shown to be constitutively and inducibly (by nafenopin and clofibrate) expressed in a sex-specific manner in male rat liver and kidney [22, 26]. Most recently, purified and reconstituted CYP4A1 was identified as a lauric acid 12-hydroxylase, while a reconstituted mixture of purified CYP4A2 and CYP4A3 efficiently catalyzed lauric acid 11- and 12-hydroxylation, yielding a product ratio of 1:3. Additionally, CYP4A2 and/or CYP4A3 were identified as testosterone 7α-hydroxylases, while CYP4A1 did not express any catalytic activity with this model substrate [27].

Immunoblotting has become a widespread technique to study the expression of numerous proteins. In experimental toxicology, immunoblotting has been employed as a sensitive tool to monitor constitutive expression and foreign compound-related changes in the expression and regulation of various phase I and phase II drug metabolizing enzymes.

Most recently, monoclonal antibodies against the peroxisome-proliferator regulated isoenzymes of the CYP4A subfamily have been successfully employed in the elucidation of their sex- and tissue specific expression, extent of inducibility and characterization of their catalytic properties [21, 22, 27].

In the following, the immunoblotting technique will be introduced [2, 4, 5, 7, 8, 12, 14, 16, 17, 18, 19, 23, 25, 28], and examples will be given for its use in:

- the assessment of cytochrome P450 induction (CYP1A, CYP2B, CYP4A) in experimental toxicology in general,
- the assessment of peroxisome proliferator-induced CYP4A isoenzyme expression,
- an approach for quantitative immunoblot analysis.

References

1. Aoyama T, Hardwick JP, Imaoka S, Funae Y, Gelboin HV, Gonzalez FJ (1990) Clofibrate-inducible rat hepatic P450s IVA1 and IVA3 catalyze the ω- and $(\omega$-1)-hydroxylation of fatty acids and the ω-hydroxylation of prostaglandins E_1 and $F_{2\alpha}$. J Lipid Res 31:1477-1482

2. Beisiegel U (1986) Protein blotting (Review). Electrophoresis 7:1-18

3. Burke MD, Thompson S, Elcombe CR, Halpert J, Haaparanta T, Mayer RT (1985) Ethoxy-, pentoxy- and benzyloxy-phenoxazones and homologues: a series of substrates to distinguish between different induced cytochromes P-450. Biochem Pharmacol 34:3337-3345

4. Gershoni JM, Palade GE (1983) Review: Protein Blotting: Principles and applications. Anal Biochem 131:1-15

5. Goding JW (1986) Monoclonal Antibodies: Principles and Practice, Second Edition, Academic Press, London

6. Gonzales FJ (1989) The molecular biology of cytochrome P450s. Pharmacol Rev 40:243-288

7. Harlow E, Lane D (1988) Antibodies: A Laboratory Manual. Cold Spring Harbor Laboratory

8. Kaufmann SH, Ewing CM, Shaper JH (1987) The erasable Western blot. Anal Biochem 161:89-95

9. Kimura S, Hanioka N, Matsunaga E, Gonzalez FJ (1989) The rat clofibrate-inducible CYP4A gene subfamily I: Complete intron and exon sequence of the CYP4A1 and CYP4A2 genes, unique exon organization, and identification of a conserved 19bp upstream element. DNA 8:503-516

10. Kimura S, Hardwick JP, Kozak CA, Gonzalez FJ (1989) The rat clofibrate-inducible CYP4A subfamily II: cDNA sequence of IVA3, mapping of the Cyp4a Locus to mouse chromosome 4, and coordinate and tissue-specific regulation of the CYP4A genes. DNA 8:517-525

11. Laemmli UK (1970) Cleavage of structural proteins during the assembly of the head of bacteriophage T4. Nature 227:680-685

12. Matthaei S, Baly DL, Horuk R (1986) Rapid and effective transfer of integral membrane proteins from isoelectric focusing gels to nitrocellulose membranes. Anal Biochem 157:123-128

13. Nelson DR, Kamataki T, Waxman DJ, Guengerich FP, Estabrook RW, Feyereisen R, Gonzalez FJ, Coon MJ, Gunsalus IC, Gotoh O, Okuda K, Nebert DW(1993) The P450 superfamily: update on new sequences, gene mapping, accession numbers, early trivial names of enzymes, and nomenclature. DNA Cell Biol 12:1-51

14. Ohlsson BG, Weström BR, Karlsson BW (1987) Enzymoblotting: Visualization of electrophoretically separated enzymes on nitrocellulose membranes using specific substrates. Electrophoresis 8:415-420

15. Okey, AB (1990) Enzyme induction in the cytochrome P-450 system. Pharmac Ther 45:241-298

16. Otey CA, Kalnoski MH, Bulinski JC (1986) A procedure for the immunoblotting of proteins separated on isoelectric focusing gels. Anal Biochem 157:71-76

17. Otter T, King SM, Witman GB (1987) A two-step procedure for efficient electrotransfer of both high-molecular weight (> 400,000) and low-molecular-weight (> 20,000) proteins. Anal Biochem 162:370-377

18. Peters JH, Baumgartner H, Schulze M (1985) Monoklonale Antikoerper: Herstellung und Charakterisierung, Springer-Verlag, Berlin

19. Pluskal MG, Przekop MB, Kavonian MR, Vecoli C, Hicks DA (1986) Immobilon PVDF transfer membrane: a new membrane substrate for Western blotting of proteins. Biotechniques 4:272-282

20. Ryan DE, Levin W (1990) Purification and characterization of hepatic microsomal cytochrome P-450. Pharmac Ther 45:153-239

21. Savoy C (1991) Monoclonal antibodies diagnostic for individual members of the cytochrome P-450IV gene family: valuable tools for studies on constitutive and nafenopin-inducible expression in Sprague Dawley rat liver, kidney and lung. Ph.D. Thesis, University of Basel, Switzerland

22. Savoy C, Wolf CR, Villermain M, Thomas H, Waechter F (1990) Monoclonal antibodies diagnostic for individual members of the cytochrome P-450IV gene family. VIIIth International Symposium on Microsomes and Drug Oxidations, Stockholm, Abstract No. 205

23. Small GM, Imanaka T, Lazarow PB (1988) Immunoblotting of hydrophobic integral membrane proteins. Anal Biochem 169:405-409

24. Sonderfan AJ, Arlotto MP, Dutton DR, McMillen SK, Parkinson A (1987) Regulation of testosterone hydroxylation by rat liver microsomal cytochrome P-450. Arch Biochem Biophys 255:27-41

25. Stott DI (1988) Immunoblotting and dot blotting (Review). J Immunol Meth 119:153-187
26. Sundseth SS, Waxman DJ (1992) Sex-dependent expression and clofibrate inducibility of cytochrome P-4504A fatty acid ω-hydroxylases: male specificity of liver and kidney CYP4A2 mRNA and tissue-specific regulation by growth hormone and testosterone. J Biol Chem 267:3915-3921
27. Thomas H, Molitor E, Kuster H, Savoy C, Wolf CR, Waechter F (1992) Hydroxylation of lauric acid and testosterone by rat liver cytochrome P450IVA proteins in reconstituted systems. In: Archakov AI and Bachmanova GI, (eds) Cytochrome P-450: Biochemistry and Biophysics INCO-TNC Joint Stock Company, Moscow, pp. 174-176
28. Towbin H, Gordon J (1984) Immunoblotting - current status and outlook. J Immunol Methods 72:313-340

8.2 Quantitative Determination of Total Cytochrome P-450 in Microsomal Fractions

8.2.1 Rat Treatment

Many drug-metabolizing enzymes have been found to be inducible by different xenobiotics. Among them there are isozymes of cytochrome P450. Examples of prototype inducers are:

- 3-methylcholanthrene which induces P450 IA forms,
- phenobarbital which induces P450 IIB forms, and
- peroxisome proliferators which induce P450 IV forms.

The treatments are performed as follows:

- Male Sprague-Dawley rats (200g) are used.
- Phenobarbital is dissolved in 0.9 % saline. Doses of 80 mg/kg rat/day are given intraperitoneally (i.p.), during 3 consecutive days. Rats are killed 24h after the last injection.
- 3-Methylcholanthrene is dissolved in corn oil and given at doses of 20mg/kg rat/day i. p. during 3 days. Animals are killed on the 4th day.
- Peroxisome proliferator, ciprofibrate, is dissolved in corn oil and given at doses of 2 mg/kg rat/day, by i. p. during 3 days. Animals are killed on the 4th day.

8.2.2 Preparation of Liver Subcellular Fractions

The experimental procedure can be summarized as follows (see Fig. 1):

- At the end of the treatment period, rats are killed by cervical dislocation. The liver is rapidly removed, weighed and perfused with buffer A (1.15 %, w/v, KCl and 50mM sodium phosphate, pH 7.4). All subsequent operations are carried out at 4°C.
- The liver is then immediatly homogenized in buffer A using four up-and-down strokes of a Potter Elvehjem glass Teflon homogenizer (1300 t/min). The ratio liver weight on homogenate volume is 25:75.
- This material is centrifuged at 400 g for 5 min at 4°C, and the resulting supernatant is centrifuged at 15,000 g for 15 min at 4°C.
- The pellet, which contains nuclei, lysosomes, mitochondria and peroxisomes, is homogenized in buffer B (0.25 M sucrose, 5mM EDTA, and 20 mM Tris-HCl, pH 7.4). The supernatant from the 15,000 g centrifugation is centrifuged at 105,000 g for 1 h at 4°C. The resulting pellet (microsomes) is homogenized in buffer A, washed once by centrifugation for 1 h at 105,000 g and finally resuspended in buffer C (20 %, v/v, glycerol, 0.1 mM EDTA and 100mM sodium phosphate, pH 7.4) at a final concentration of 1 g of liver/ml.
- Aliquots of peroxisomal or microsomal fractions are stored at -70°C for future assays.

8.2.3 Protein Assay

Protein is determined by the method of LOWRY et al [2].
- Prepare the following solutions:

  - standard bovine serum albumin, 1mg/ml of 0.05 M NaOH
  - 2 % (w/v) sodium potassium tartrate
  - 0.05 M sodium hydroxide
  - 1 % (w/v) copper sulphate, $5H_2O$
  - 0.05M sodium hydroxide containing 2 % (v/v) anhydrous sodium carbonate in distilled water.

- A range of standards is prepared from the stock albumine solution, 0, 25, 50, 100, 150, 200, and 250 µg/ml, in 0.05 M NaOH.
- Samples are diluted as appropriate in 0.05 M NaOH.
- Sodium hydroxide (with sodium carbonate), copper sulphate and sodium potassium tartrate solutions are mixed in the ratio 100:1:1 (by volume) immediately before use. Tartrate is added before copper sulphate.
- 5 ml of this mixture are added to 0.5 ml of the samples, standards and 0.05 M NaOH blank.

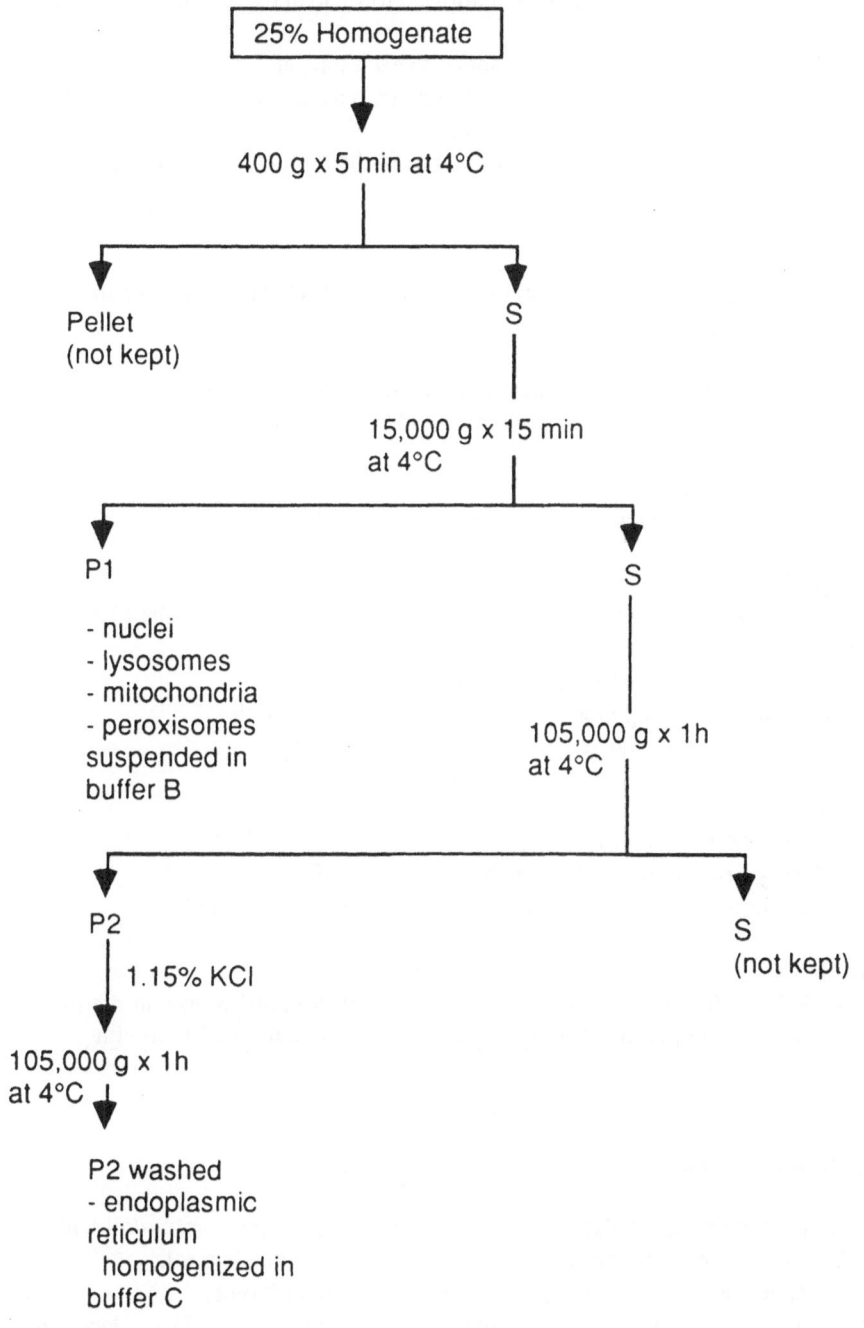

Figure 1: Preparation of subcellular liver fractions.

- The tubes are mixed immediately and allowed to stand at room temperature for 15 min.
- 0.5 ml of Folin-Ciocalteu reagent is added to each tube and mixed immediately. After 35 min, the OD is recorded at 700 nm against a distilled water blank on a spectrophotometer (Uvikon series, Kontron).
- The concentration of protein is determined from the standard curve of A_{700} versus albumin concentrations ($\mu g/ml$).

8.2.4 Quantitative Determination of Total Cytochrome P-450 in Hepatic Microsomes

Cytochrome P-450 is an hemoprotein, characterized by the presence of an absorbance band at about 450 nm for the CO-adduct of the reduced form.

8.2.4.1 Equipment, Chemicals and Solutions

Equipment
- a double beam spectrophotometer UV-visible equipped with a recorder (Uvikon series, Kontron)
- 4 ml disposable cuvettes for spectrophotometer
- adjustable automatic pipettes (Gilson or Labsystems) with tips
- 20ml disposable cups (Sarstedt)

Chemicals
- $Na_2S_2O_4$ (sodium dithionite) (1 g)
- HEPES ([hydroxyethyl(piperazinylethane)] ethanesulfonic acid) (250 g)
- a 800 cm^3 bottle of CO gas, reagent purity (Alphagaz, Air Liquide)

Solutions
- 0.1 M HEPES buffer, pH 7.4: dissolve 5.9 g HEPES in about 200 ml distilled water. Adjust the pH to 7.4 with 1N NaOH and make up to 250 ml with distilled water.

8.2.4.2 Experimental Procedure

- The microsomes are diluted to a concentration of 1-1.5 mg of protein/ml with HEPES buffer in 20 ml disposable cups. The final volume is 6 ml.
- This suspension is reduced by a few grains of sodium dithionite.
- The diluted suspension is divided equally into two cuvettes, which are placed in the cell compartment of the spectrophotometer.
- A base line is recorded between 500 and 400 nm (scan speed 100 nm/min).

- The content of the sample cuvette is gassed for about 1 min with carbon monoxide.
- The spectrum is then recorded, and the change in absorbance at 450 nm relative to 490 nm is measured and is converted to the concentration of cytochrome P-450 by use of the molar extinction coefficient 91 $cm^{-1}mM^{-1}$.

8.2.4.3 Calculation of the Cytochrome P-450 Concentration

$$\text{nmoles cytochrome P450/mg protein} = \frac{(A450-A490)}{91 \times [P]} \times 1000$$

[P] = microsomal protein concentration in the cuvette

References

1. Guengerich FP (1989) Analysis and characterization of enzymes. In: Hayes AW (ed, 2nd edition) Principles and Methods in Toxicology, Raven Press, Ltd., New York, pp 777-786
2. Lowry OH, Rosebrough NJ, Farr AL, Randall RJ (1951) Protein measurement with the Folin phenol reagent. J Biol Chem 193:265-275
3. Omura T, Sato R (1964) The carbon monoxide-binding pigment of liver microsomes. Evidence for its hemoprotein nature. J Biol Chem 239:2370-2378

8.3 Induced Enzymatic Activity Assays

8.3.1 Ethoxyresorufin O-Deethylase (EROD) and Pentoxyresorufin O-Depentylase (PROD)

EROD is an activity highly specific for the cytochrome P450 IA forms, preferentially induced by 3-methylcholanthrene.

PROD is an enzyme activity specific for cytochrome IIB forms which are induced by phenobarbital.

Ethoxyresorufin or pentoxyresorufin are O-dealkylated to resorufin by liver microsomes in the presence of NADPH and O_2. The reaction involves cytochrome P-450 and is monitored directly by recording the increase in fluorescence associated with the formation of resorufin (Fig. 2).

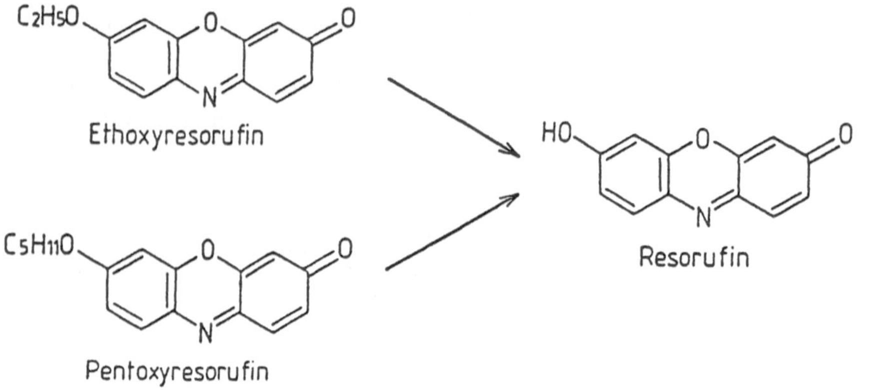

Figure 2 : De-alkylation of alkoxyresorufins.

8.3.1.1 Equipment, Chemicals and Solutions

Equipment
- a fluorimeter equipped with thermostable cell holder (Uvikon SFM 25, Kontron) and a water bath
- 4 ml disposable acryl cuvettes for fluorimeter (Sarstedt)
- a chronometer
- adjustable automatic pipettes (Gilson or Labsystems) with tips

Chemicals
- 7-ethoxyresorufin (>95%, Boehringer, 5 mg)
- 7-pentoxyresorufin (>95%, Boehringer, 10 mg)
- resorufin (Aldrich, 5 g)
- NADPH tetrasodium salt (Sigma, 100 mg)
- Tris (100 g)
- $MgCl_2$, $6H_2O$ (250 g)
- methanol

Solutions
- 0.1 M Tris-HCl buffer pH 7.7: dissolve 1.21 g Tris in about 50 ml distilled water. Adjust the pH to 7.7 with 1 N HCl and make up to 100 ml.
- 500 mM $MgCl_2$, $6H_2O$: dissolve 10.2 g in 100 ml distilled water.
- 100 μM 7-ethoxyresorufin : dissolve 3.03 mg in 25 ml methanol. Subsequently dilute this solution 5 fold in methanol.
- 1 mM 7-pentoxyresorufin: dissolve 7.1 mg in 25 ml methanol.
- 0.005 mM resorufin: dissolve 11.7 mg in 50 ml methanol. Subsequently dilute this solution 100 fold in Tris buffer.
- 25 mM NADPH: dissolve 104.2 mg in 5 ml distilled water. Store in a freezer.

8.3.1.2 Experimental Procedure

Assay Conditions
- Tris-chloride buffer, pH 7.7   100 mM
- $MgCl_2$ 25 mM
- Ethoxyresorufin   1 μM
   or
   Pentoxyresorufin   10 μM
- NADPH   250 μM
- Microsomal proteins   0.2-0.5 mg/ml
- Final volume   2 ml
- Excitation wavelength   522 nm
- Emission wavelength   586 nm
- Temperature   30 °C

Measurement
- The general mixture is prepared in a fluorimeter cuvette as follows:

  - 20 μl of ethoxyresorufin or pentoxyresorufin
  - 1840 μl of Tris-chloride buffer
  - 100 μl of $MgCl_2$ solution
  - 20 μl of microsome suspension

- The excitation and emission wavelengths are adjusted to 522 nm and 586 nm, respectively.
- The mixture is allowed to stand at 30°C during 3 min and the reaction is started by adding 20 μl of 25 mM NADPH.
- The progressive increase of fluorescence is recorded during 3-4 min. The fluorimeter is then calibrated with 10 μl of 0.005 mM resorufin (50 pmoles).

8.3.1.3 Calculation of Enzyme Activity

The rate of formation of resorufin is calculated by comparing the increase of fluorescence during 1 min to the fluorescence of known amounts of resorufin.

Activity is expressed as pmoles of resorufin formed/min/mg microsomal protein:

$$\text{Activity} = \frac{100 \times F}{ST \times [P]}$$

F: increase of fluorescence during 1 min
ST: fluorescence of 100 pmoles of resorufin
[P]: amount (mg) of microsomal protein in the cuvette

8.3.2 Peroxisomal Acyl-CoA Oxidase

Peroxisomal β-oxidation is determined as the palmitoyl-CoA-dependent reduction of NAD^+ in the presence of cyanide to inhibit reoxidation of NADH by mitochondria, as described by Lazarow [2], with some modifications according to Mitchell [4].

8.3.2.1 Equipment, Chemicals and Solutions

Equipment
- a spectrophotometer (Kontron series Uvikon)
- a balance (to 1/10 mg)
- quartz cuvettes
- a vortex mixer
- a timer
- adjustable automatic pipettes (Gilson or Labsystems) and tips
- ice and a storage container
- glassware
- a squeeze bottle for distilled water
- filter paper

Chemicals
- fatty acid free bovine serum albumin, fraction V (Boehringer, 1 g)
- dithiothreitol (Sigma, 2 g)
- niacinamide (nicotinamide; Sigma, 100 g)
- palmitoyl CoA (Pharmacia, 25 mg)
- flavin adenine dinucleotide (FAD, Boehringer, 100 mg)
- β-nicotinamide adenine dinucleotide (NAD; Sigma, 100 mg)
- coenzyme A (sodium salt, Sigma, 25 mg)
- potassium cyanide (100 mg)
- Triton X-100 (100 ml)
- Tris (hydroxymethyl)-aminomethane (100 g)

Solutions
- 60 mM Tris-HCl, pH 8.3: dissolve 3.63 g of Tris in 500 ml H_2O. Adjust the pH to 8.3 with HCl.
- Reagent A: 75 µM coenzyme A, 555 µM NAD, 4.2 mM dithiothreitol, 3 mM potassium cyanide, 0.0225 % (w/v) bovine albumine, and 141 mM nicotinamide
- flavin adenine dinucleotide, 14.9 mg/ml in 60 mM Tris HCl buffer
- palmitoyl-CoA, 7.5 mg/ml in 60 mM Tris HCl buffer
- 1% (v/v) Triton X-100: 1 ml/100 ml in 60 mM Tris HCl buffer

8.3.2.2 Experimental Procedure

- Samples are diluted 1:1 with 1 % Triton X-100 and incubated at 37°C for 5 min prior to assay. Reagent A (2 ml), FAD solution (20 µl) and an aliquot of pellet P_1 (containing peroxisomes, see Fig 1) at the final protein concentration <0.2 mg/ml are added to each of two cuvettes.
- The volume is adjusted to 3 ml with 60 mM Tris-HCl. The components are mixed and equilibrated at 37°C.
- A baseline is recorded at 340 nm.
- Following addition of palmitoyl-CoA solution (20 µl) to the sample cuvette, the time-dependent DO at 340 nm is recorded.
- Peroxisomal β–oxidation is expressed as nmol NAD^+ reduced min^{-1} mg $protein^{-1}$ using an extinction coefficient of 6.22 mM^{-1} cm^{-1} for NADH.

8.3.3 Lauric Acid Hydroxylase

8.3.3.1 Equipment, Chemicals and Solutions

Equipment
- a HPLC pump (Kontron 420)
- a radioactivity detector (Trace 7140, Packard)
- an integrator (Microplot 44T, Gulton)
- a 4.6 mm x 150 mm, 5µm Nucleosil C_{18} column (Interchrom, Interchim)
- a rotator for tubes
- a freezer (-40°C)
- a 25 µl microsyringe (Hamilton)
- a N_2 gas tank connected to a series of Pasteur pipettes
- a low speed centrifuge (Beckman, GS6-KR) and a rotor equipped for 10 ml tubes
- a thermostated water bath (37°C)
- 10 ml glass tubes with screw caps and racks
- Pasteur pipettes + bulbs
- a timer
- a vortex mixer
- adjustable automatic pipettes (Gilson or Labsystems) and tips
- ice and a storage container
- 2 ml glass vials with screw caps
- 50 ml beakers

Chemicals
- β-nicotinamide adenine dinucleotide phosphate (NADPH; Sigma, 100 mg)
- lauric acid, sodium salt (Sigma, 25 mg)
- [1-^{14}C] lauric acid (52 mCi/mmol, Amersham, 250 µCi in 2.5 ml hexane)
- ammonium acetate

- acetonitrile for HPLC
- methanol
- ultrapure water
- diethyl ether
- hydrochloric acid
- Tris (hydroxymethyl)-aminomethane (Tris)

Solutions
- 66 mM Tris HCl buffer, pH 7.4: dissolve 4 g Tris in 500 ml H_2O. Adjust the pH to 7.4 with HCl.
- 1.5 mM lauric acid (sodium salt) in Tris HCl pH 7.4
- 1.5 mM labelled lauric acid: dry down 110µl (11 µCi) of [1-^{14}C] lauric acid under nitrogen and add 10 ml of 1.5 mM unlabelled lauric acid
- 25 mg/ml NADPH in 66 mM Tris HCl buffer

HPLC Eluents
- eluent A: 315 ml acetonitrile + 415 ml ultrapure water + 270 ml 0.1 M ammonium acetate, pH 4.6. The 0.1 M ammonium acetate solution is prepared by dissolving 7.7 g NH_4OOCCH_3 in 1 L ultrapure water; adjust the pH to 4.6 with CH_3COOH. Filter eluent A using a vacuum filter (Millipore).
- eluent B: 100% acetonitrile

8.3.3.2 Experimental Procedure

Preparation of samples
- Microsomal suspension (0.2-1mg protein), 1.5 mM [1- ^{14}C] lauric acid (170 µl) and Tris-HCl buffer are placed in 10 ml stoppered glass test tubes to give a final volume of 1900 µl.
- The tubes are preincubated at 37°C for 3 min. The reaction is initiated by addition of 100 µl of the NADPH solution.
- The incubation time at 37°C depends on the metabolism extent expected from the sample (5-15 min).
- After this incubation time, the reaction is stopped by addition of HCl (500 µl), and the tubes are kept on ice.
- [1- ^{14}C] lauric acid and its [1- ^{14}C] hydroxylated products are extracted with 2 ml diethyl ether (15 min), twice.
- The ether layer is aspirated and dried down under nitrogen. At this step, samples can be kept at -20°C. Lauric acid and its metabolites are resuspended in methanol (20 µl).
- 11-Hydroxy lauric acid or 12-hydroxy lauric acid and lauric acid are separated by high pressure liquid chromatography (HPLC).

HPLC Analysis

- 11- and 12-hydroxy lauric acid are separated on the C^{18} column with eluent A at a flow rate of 1.5 ml/min. After these have been eluted, lauric acid is eluted with 100 % acetonitrile (see Fig 3).
- [1-^{14}C] lauric acid and its [1-^{14}C] metabolites are detected with the Packard radioactivity detector, and the metabolites are quantitated by determination of peak area as percentages of the [1-^{14}C] laurate peak area using the Gulton integrator.
- Results are expressed as nmoles produced metabolites/min/mg proteins.

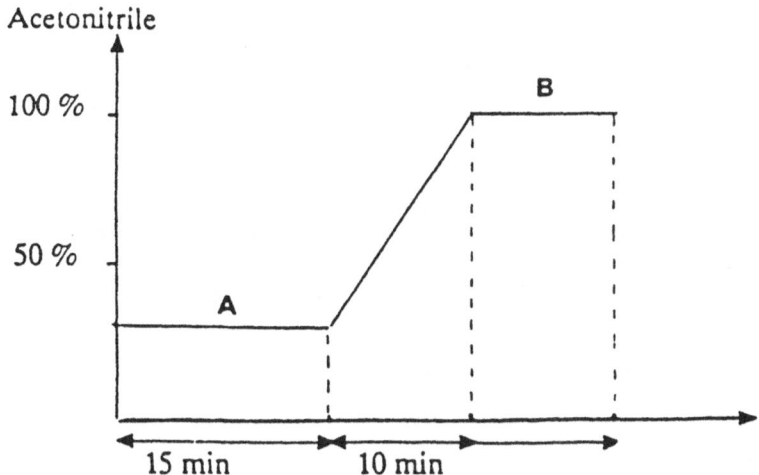

Figure 2: Scheme of the acetonitrile gradient used for HPLC analysis of lauric acid and its metabolites.

References

1. Burke MD, Mayer RT (1974) Ethoxyresorufin: a direct fluorimetric assay of a microsomal O-dealkylation which is preferentially inducible by 3-methylcholanthrene. Drug Metabolism and Disposition 2:583-588
2. Lazarow PB (1981) Assay of peroxisomal b-oxidation of fatty acids. In Lowenstein JM (ed.) "Methods in Enzymology," vol. 72, lipids Part D, Academic Press, New York, pp. 315-317
3. Lubet RA, Mayer RT, Cameron JW, Nims RW, Burke MD, Wolf T, Guengerich FP (1985) Dealkylation of pentoxyresorufin: a rapid and sensitive assay for measuring induction of cytochrome(s) P450 by phenobarbital and other xenobiotics in the rat. Archives of Biochemistry and Biophysics 238:43-48
4. Mitchell AM (1985) Hepatic peroxisome proliferation: mechanisms and species differences. Doctoral thesis, University of Surrey, Guildford, England

5. Parker GL, Orton TC (1980) Biochemistry, Biophysics and Regulation of Cytochrome P450. In Gustaffson JA, Duke JC, Mode A and Rafter J (Eds) Elsevier, North Holland-Press, Amsterdam, pp. 373-377

8.4 Immunoblotting Analysis of Cytochrome P-450 Expression in Rat Liver

H. THOMAS, M. VILLERMAIN and F. WAECHTER

8.4.1 Equipment, Chemicals and Solutions

8.4.1.1 Equipment

Access to
- a vertical mini slab gel electrophoresis cell (Mini-PROTEAN[TM], Bio-Rad Laboratories)
- a mini electrophoretic transfer cell (Mini Trans-Blot[TM] system, Bio-Rad Laboratories)
- an electrophoresis power supply
- a refrigerator (4°C)
- a deep freezer (-80°C)
- deionized water
- double distilled water
- vacuum

On the Bench
- a vortex mixer
- a magnetic stirrer with Teflon-coated magnetic rods
- a pH meter
- disposable examination gloves
- glassware
- disposable plastic pipettes (5 and 10 ml)
- adjustable automatic pipettes (Gilson or Labsystems) and tips
- 50 ml Falcon tubes
- 1.5 ml Eppendorf reaction tubes
- sterile plastic Petri dishes (14 cm in diameter)
- a pair of scissors or a cutter for nitrocellulose
- a ruler, 30-50 cm in length
- a pair of flat-nosed plastic tweezers
- a timer
- a water bath (up to 100°C) with racks to hold 1.5 ml Eppendorf tubes
- 10 ml plastic syringes equipped with 21G1.5 needles

- Hamilton syringes (10-50 μl)
- a shaker with adjustable frequency
- Whatman filter paper
- nitrocellulose transfer membrane 0.45 μm (Bio-Rad No. 162-0115)
- a glass funnel (15 cm in diameter) and filter paper
- plastic trays with lids (19 x 14 x 6 cm)
- Pasteur pipettes with rubber bulbs
- a mini-PROTEAN II, Cell Assembly Guide

8.4.1.2 Chemicals and Biological Material

- Tris base
- glycine
- sodium dodecylsulfate (SDS)
- ammonium persulfate
- acrylamide (electrophoresis grade)
- bisacrylamide (electrophoresis grade)
- N,N,N',N'-tetramethylethylenediamine (TEMED; electrophoresis grade)
- methanol
- glycerol
- 2-mercaptoethanol
- 30% hydrogen peroxide
- bromophenol blue
- 2 N hydrochloric acid
- 1 N hydrochloric acid
- phosphate buffered saline (PBS): Dulbecco's PBS without Ca^{2+} and Mg^{2+}
- horse serum (heat inactivated, Gibco-BRL)
- 4-chloro-1-naphthol (chloronaphthol)
- isobutanol
- microsomal suspensions from untreated, phenobarbitone (PB)-, 3-methylcholanthrene (3-MC)-, and ciprofibrate (CIP)-treated male Sprague Dawley rat liver, stored at -80°C
- purified CYP4A1 and a mixture of purified CYP4A2/4A3
- monoclonal antibodies (purified mouse IgG) be4, clo1, clo6 and d8, as stock solutions containing 1 mg IgG/ml and stored at - 80°C
- horseradish peroxidase conjugated sheep anti-mouse IgG, A, M (H and L chains; Binding Site Code No. PP270)

8.4.1.3 Solutions

Separating and Stacking Gels
- 30% (w/v) acrylamide + 0.8% (w/v) bisacrylamide in double distilled water
- 2 M Tris/HCl buffer, pH 8.8
- 0.5 M Tris/HCl buffer, pH 6.8
- 10% (w/v) ammonium persulfate in double distilled water (prepared immediately prior to use).
- 10% (w/v) SDS in double distilled water.
- water-saturated isobutanol.

Sample Buffer
Dissolve 2 g SDS in 16 ml 0.5 M Tris/HCl buffer pH 6.8, add 10 ml glycerol, 5 ml 2-mercaptoethanol and 30 mg bromophenol blue. Make up to 100 ml with double distilled water.

Running Buffer
Dissolve 3.0 g Tris base, 1.0 g SDS and 14.4 g glycine in 1 L double distilled water.

Blotting Buffer
Dissolve 3.0 g Tris base and 14.4 g glycine in about 500 ml double distilled water, add 200 ml methanol and adjust to 1 L with double distilled water.

Blocking Buffer
Prepare 10% (v/v) horse serum in PBS.

Staining Solution
Prepare freshly immediately prior to use by mixing 47 ml PBS, 3 ml 0.3% (w/v) chloronaphthol in methanol and 25 µl 30% hydrogen peroxide.

8.4.2 Experimental Procedure

8.4.2.1 SDS Polyacrylamide Gel Electrophoresis (SDS-PAGE)

1. Casting Discontinuous Gels
1.1. Assembling the Glass Plate Sandwiches

Note: To insure proper alignment and casting, make sure plates, spacers, combs, and casting stand gaskets are clean and dry before proceeding. Details on the assembly of the electrophoresis cell will be provided in the Bio-Rad "Mini-PROTEAN, II Cell Assembly Guide," which will be available at the bench.

[1] Assemble the gel sandwich on a clean surface. Lay the longer rectangular glass plate down first, then place the two spacers of equal thickness along the short edges of the rectangular plate. Next place the shorter glass plate on top of the spacers so that the bottom end of the spacers and glass plates are aligned. At this point, the spacers should be sticking up above the long glass plate by about 5 mm.

[2] Loosen the four screws on the clamp assembly and stand it up so that the screws are facing away from you. Firmly grasp the glass plate sandwich with the tall plate facing away from you and gently slide it into the clamp assembly along the front face of the acrylic pressure plate. The longer glass plate should be up against the acrylic pressure plate of the clamp assembly. Tighten the top two screws of the clamp assembly.

[3] Using the leveling bubble, level the casting stand so that the alignment bar is facing you. Check to see that the removable silicone gaskets are clean and free of residual acrylamide to insure a good seal. Place the silicone rubber gasket on top of the permanent foam pads of the casting stand slots.

[4] Place the clamp assembly up against the alignment bar so that the clamp screws face away from you. Loosen the top two screws to allow the plates and spacers to set flush against the casting stand base. Check that the spacers are flush against the sides of the clamps. If not, relocate them properly using the top portion of the spacer for adjustment. Then recheck that the plates and spacers are flush to the casting stand base. While holding the clamps firmly against the glass plates with one hand, re-tighten the top two clamp screws.

[5] Remove the clamp assembly from the alignment bar and tighten the bottom two clamp screws. The sandwich is now properly aligned and ready for gel casting.

[6] Transfer the clamp assembly to one of the casting slots of the casting stand. If two gels are to be cast, place the clamp assembly on the side opposite of the alignment bar to make aligning the next sandwich easier.

[7] Attach the sandwich in the following way: Butt the acrylic pressure plate against the wall of the casting slot at the bottom, so that the glass plates rest on the rubber gasket. Snap the acrylic plate underneath the overhang of the casting slot by pushing with the white portions of the clamps (Do not push against the glass plates or spacers; this could cause breakage).

1.2. Casting Discontinuous (Laemmli) Gels

Discontinuous gels consist of a resolving or separating (lower) gel and a stacking (upper) gel. The stacking gel acts to concentrate large sample volumes, resulting in better band resolution than is possible using the same volumes on a gel without a stack. Molecules are then completely separated in the resolving gel. The most popular discontinuous buffer system is that of Laemmli [1] given below.

[1] Prepare a 10 % separating gel monomer solution (sufficient for 2 gels) by combining the following reagents in a 50 ml flask:

- 3.70 ml of 30 % acrylamide / 0.8% bisacrylamide,
- 2.90 ml of 2 M Tris/HCl buffer, pH 8.8,
- 0.12 ml of 10% SDS and
- 4.20 ml of double distilled water.

  Deaerate the solution under vacuum for about 15 min.

[2] Place a comb completely into the assembled gel sandwich. With a marker pen, place a mark on the glass plate 1 cm below the teeth of the comb. This will be the level to which the separating gel is poured. Remove the comb.

[3] Add 0.08 ml of 10 % ammonium persulfate and 4 µl TEMED to the deaerated monomer solution and pour the solution to the mark, using a Pasteur pipette and a bulb. The easiest way to pour is to flow the solution down the middle of the longer plate of the gel sandwich. Another way to pour is to flow the solution down the side of one of the spacers. Pour the solution smoothly to prevent it from mixing with air.

[4] Immediately overlay the monomer solution with water-saturated isobutanol. The advantage of using isobutanol is that the overlay solution can be applied rapidly with a Pasteur pipet and bulb because very little mixing will occur. If water is used to overlay, it must be done using a needle and syringe, using a steady, even rate of delivery to prevent mixing.

[5] Allow the gel to polymerize for 1 hour. Rinse off the overlay solution completely with distilled water. This is especially important with an alcohol overlay. Do not allow alcohol to remain on the gel more than 1 h, or dehydration of the top of the gel will occur.

[6] Prepare a 2% stacking gel monomer solution (sufficient for 2 gels). Mix the following reagents in a 25 ml flask:

- 0.75 ml of 30% acrylamide /0.8% bisacrylamide,
- 1.00 ml of 0.5 M Tris/HCl buffer, pH 6.8,
- 0.04 ml of 10% SDS and
- 2.60 ml of double distilled water.

Deacrate under vacuum for about 15 min.

[7] Dry the area above the separating gel with filter paper before pouring the stacking gel.

[8] Place a comb in the gel sandwich and tilt it so that the teeth are at a slight (~10°) angle. This will prevent air from being trapped under the comb teeth while pouring the monomer solution.

[9] Add 0.04 ml of 10 % ammonium persulfate and 4 µl TEMED to the degassed monomer solution and pour the solution down the spacer nearest the upturned side of the comb. Pour until all the teeth have been covered by solution. Then properly align the comb in the sandwich and add monomer to fill completely. The comb is properly seated when the T portion of the comb rests on top of the spacers. Generally, an overlay solution is not necessary for polymerization when a comb is in place.

[10] Allow the gel to polymerize for 30 min. Remove the comb by pulling it straight up slowly and gently.

[11] Rinse the wells completely with distilled water. The gels are now ready to be attached to the inner cooling core, the sample loaded, and the gels run.

2. Sample Preparation

[1] Prepare dilutions (1 ml) of microsomal liver fractions from PB-, 3-MC- and CIP-treated rats with sample buffer in 1.5 ml Eppendorf reaction tubes to obtain final concentrations of 5, 40 and 200 µg microsomal protein/ml. Microsomal fractions from untreated animals are adjusted with sample buffer to a final protein concentration of 2.0 mg/ml. These protein concentrations were selected on the basis of our laboratory experience and may be changed depending on the liver enzyme induction obtained with the animals to be used in this experiment.

[2] Heat the samples for 5 min in a boiling water bath. Subsequently, allow the samples to cool down to room temperature. They are now ready to use.

3. Assembling the Upper Buffer Chamber

Note: To insure a leakproof seal, make sure the U-shaped inner cooling core gaskets are clean and that they are installed with the notched side exposed for contact with the gel sandwich.

[1] Release the clamp assemblies/gel sandwiches from the casting stand.

[2] Lay the inner cooling core down flat on a lab bench. With the glass plates of the clamp assembly facing the cooling core (and the clamp screws facing up), carefully slide the clamp assembly wedges underneath the locator slots on the inner cooling core until the inner glass plate of the gel sandwich butts up against the notch in the U-shaped gasket.

Note: If the clamp is introduced at too steep an angle, the inner glass plate will pinch the notched gasket and leaking will result. Lubricate the raised portions of the gasket with a drop of running buffer or water to help the glass plate sandwich slide in properly. It is extremely important that the short glass plate is firmly seated against the notch in the gasket, or buffer leaks will occur. Visually inspect the glass plate/notch interface to insure correct assembly.

[3] When the inner plate is in position, push down on the base of the clamp assembly until the latch engages each side.

[4] Turn over the electrode assembly and attach another clamp assembly to the other side in the same manner.

4. Loading the Samples

Sample loading can be done in two ways. The most common method is to load samples into wells formed in the gel with a well-forming comb. The second method (not performed in this course) uses the entire gel surface as a single well for liquid samples .
   The sample volume that each well will hold is dependent on the gel thickness and amounts, *e.g.*, to approximately 35 µl for a 10-well comb at a gel thickness of 0.75 mm used in this experiment.

[1] Lower the inner cooling core into the lower buffer chamber of the Mini-PROTEAN II cell. Add approximately 115 ml of running buffer to the upper buffer chamber. Fill until the buffer reaches a level no higher than 2 mm from the top of the outer long glass plate.

Note: Overfilling the upper buffer chamber could result in siphoning of the buffer over the tops of the spacers, resulting in excessive buffer loss.

[2] Pour about 180 ml of the buffer into the lower buffer chamber so that at least the bottom 1 cm of the gel is covered. Remove any air bubbles from the bottom of the gel so that good electrical contact is achieved. This can be done by swirling the lower buffer with a pipet until the bubbles clear.

[3] Load the samples into the wells under the electrode buffer with a Hamilton syringe (Tables 1 and 2). Insert the syringe to about 1-2 mm from the well bottombefore delivery. Disposable pieces of plastic tubing may be attached to the syringe to eliminate the need for rinsing the syringe between samples.

Note: The sample buffer must contain either 10% sucrose or glycerol in order to underlay the sample in the well without mixing.

5. Running the Gels

The gel is run at a constant voltage of 200 volts until the bromophenol blue front reaches a 5 mm distance from the bottom of the gel.

6. Removing the Gels

[1] After electrophoresis is complete, turn off the power supply and disconnect the electrical leads.

[2] Remove the cell lid and carefully pull the inner cooling core out of the lower chamber. Pour off the upper buffer.

[3] Lay the inner cooling core on its side and remove the clamp assembly in the following manner: push down on both sides of the cooling core latch and up on the clamps until the clamp assembly is released. Slide the clamp assembly away from the cooling core.

[4] Loosen all four screws of the clamp assembly and remove the glass plate sandwich from it.

[5] Push one of the spacers of the sandwich out to the side of the plates without removing it.

[6] Gently twist the spacer so that the upper glass plate pulls away from the gel.

[7] Remove the gel by gently grasping two corners of the gel and lifting off. Place the gel for 10 min into blotting buffer.

Table 1: Distribution of gels and samples for qualitative Western-blotting

Well Number	1	2	3	4	5	6	7	8	9
Microsomal protein					(µg/well)				
Gel 1	-	-	C (20)	3-MC (1.0)	CYP2B1/2B 2 (0.1)	PB (0.2)	PB (0.8)	PB (2.0)	CIP (0.6)
Gel 2	-	-	C (20)	PB (2.0)	CYP1A1/1A 2 (0.04)	3-MC (0.1)	3-MC (0.4)	3-MC (1.0)	CIP (0.6)
Gel 3	C (20)	3-MC (20)	CYP4A1 (0.06)	CIP (0.06)	CIP (0.24)	CIP (0.6)	CYP4A2/4A 3 (0.06)	PB (2.0)	PB (2.0)
Gel 4	C (20)	3-MC (10)	CYP4A2/4A 3 (0.06)	CIP (0.2)	CIP (0.6)	CIP (1.2)	CYP4A1 (0.06)	PB (2.0)	PB (10)

Microsomal protein will be derived from untreated (C), phenobarbitone (PB)-, 3-methylcholanthrene (3-MC)-, and ciprofibrate (CIP)- treated male rat liver.

Table 2: Distribution of gels and samples for quantitative Western-blotting using purified rat liver CYP4A1 and CYP4A2/4A3 as well as microsomal liver fractions from ciprofibrate-treated male rats.

Well Number	1	2	3	4	5	6	7	8	9
Protein					(µg/well)				
Gel 5 (microsomal fractions)	-	0.03	0.1	0.2	0.5	1.0	2.5	5.0	10
Gel 6 (purified CYP4A1)	-	0.005	0.0075	0.01	0.015	0.025	0.03	0.06	0.09
Gel 7 (microsomal fractions)	-	0.01	0.02	0.05	0.1	0.25	0.5	1.0	2.0
Gel 8 (purified CYP4A2/4A3)	-	0.002	0.005	0.01	0.03	0.06	0.09	0.12	0.15

8.4.2.2 *Immunoblotting*

Note: Wear gloves for this procedure to avoid contamination of nitrocellulose membranes.

1. Equilibration of Nitrocellulose

Cut a sheet of nitrocellulose in the size of the SDS-gel and mark the position of the protein lanes to be blotted onto the cellulose with a ball-point pen. The nitrocellulose sheet is submerged in blotting buffer and allowed to equilibrate for 20 min.

2. Assembling the Gel Holder Cassette

[1] Open the gel holder cassette by sliding and lifting the latch. Note that one panel of the cassette is clear and the other panel is tinted smokey gray. The clear panel is the anode (+) side and the gray panel is the cathode (-) side. For the electrode module, the cathode is the smokey gray electrode panel located in the center of the buffer tank. Always insert the gel cassette so that the gray plastic faces the gray plastic of the cathode electrode. The anode and cathode are identified by the black (cathode) and red (anode) disks on top of the electrode module.

[2] Place the opened gel holder in a shallow vessel (e.g. plastic tray) so that the gray panel is flat on the bottom of the vessel. The clear panel should rest at an angle against the wall of the vessel.

[3] Place a pre-soaked fiber pad on the gray panel of the cassette. When assembling the fiber pads, filter paper, gel, and membrane, be sure to center all components. If any of the material extends past the edge of the cassette, it will catch on the guide rails, causing a distortion of both the blot-gel contact and the transfer pattern.

[4] Place a piece of the saturated filter paper on top of the fiber pad. Saturate the surface of the filter paper with 2-3 ml of buffer. Place the equilibrated gel on top of the paper. Align the gel in the center of the cassette. Transfer will be incomplete for any portion of the gel that is outside the pattern of circles of the gel holder cassette. Make sure that no air bubbles are trapped between the gel and the filter paper.

[5] Flood the surface of the gel with blotting buffer and lower the pre-wetted blotting media on top of the gel. This is best done by holding the membrane at opposite ends so that the center portion will contact the gel first. Then, gradually

lower the ends. Next, roll a glass pipette or test tube over the top of the membrane (like a rolling pin) to exclude all air bubbles from the area between the gel and membrane.

[6] Flood the surface of the membrane with buffer. Complete the sandwich by placing a piece of saturated filter paper on top of the membrane and placing a saturated fiber pad on top of the filter paper.

[7] Close the cassette. Hold it firmly so the sandwich will not move and secure the latch. At this point, movement of the sandwich might disrupt the gel-membrane contact which would cause incomplete transfers or swirling transfer patterns.

[8] Place the gel holder in the buffer tank so that the gray panel of the holder is facing the gray cathode electrode panel.

[9] If two gels are to be transferred, repeat steps 1-8 with the second gel. When the gels to be transferred are in place, set the buffer tank on top of a magnetic stirrer. Fill the tank with blotting buffer to just above the level of the top row of circles on the gel holder cassette. Do not overfill as buffer will leak down the sides of the unit. Turn on the magnetic stirrer, and put the lid in place. Be sure that the electrode wires on the lid are attached to the appropriate pins of the electrode module (black wire to cathode and red wire to anode). Plug the unit into the power supply. Normal transfer polarity is cathode to anode, *i.e.* red wire to red outlet and black wire to black outlet on the power supply.

3. Electrotransfer of the Separated Proteins from the Gel onto Nitrocellulose

Electrotransfer is accomplished at a constant voltage of 100 volts during 1 h.

4. Immunoassay and Detection

The procedure below relates to the processing of 1 nitrocellulose sheet. For the processing of several gels at the same time, increase the volumes of washing buffer correspondingly.

[1] Remove the nitrocellulose membrane and place it into a Petri dish containing 25 ml PBS for 15 min with continuous shaking.

[2] Transfer the membrane into another Petri dish containing 25 ml blocking buffer and allow the blocking of all protein binding sites for 2 h with gentle shaking.

[3] Incubate with blocking buffer containing the first antibody (monoclonal IgG, Table 3) at a final concentration of 1 µg/ml overnight with gentle shaking.

Table 3: Gels (blots) and monoclonal antibodies (MAb) to be used for cytochrome P-450 detection.

Gel (blot) Number	MAb	Raised against and specific/ diagnostic for rat liver
1	be4	CP2B1/2B2
2	d8	CYP1A1/1A2
3	clo1	CYP4A1
4	clo6	CYP4A2/4A3
5	clo1	CYP4A1
6	clo1	CYP4A1
7	clo6	CYP4A2/4A3
8	clo6	CYP4A2/A43

[4] Wash the introcellulose membrane thoroughly with deionized water and incubate with 25 ml of the second horseradish peroxidase conjugated goat anti-mouse antibody in blocking buffer for 2 h with gentle shaking.

[5] Wash the membrane thoroughly with deionized water and place it into 25 ml of staining solution for 20 to 30 min.

[6] Stop the staining process by washing the membrane with deionized water.

[7] Rinse the membrane well and let it dry between two layers of Whatman filter paper.

5. Quantification of Cytochrome P-450 Expression

Cytochrome P-450 (CYP4A) expression may be quantified optionally by densitometric scanning with a Camag TLC scanner II using a Cats version 3 software.

[1] Scan the immunoblots 6 and 8, which contain increasing concentrations of the purified cytochromes CYP4A1 (blot 6) and CYP4A2/4A3 (blot 8), in the reflecting mode at a wavelength of 550 nm.

[2] Establish a calibration curve by plotting the signal integrals as obtained just above vs. the corresponding protein concentrations given in Table 2.

[3] Similarly plot the integrals obtained with increasing concentrations of microsomal protein and MAb clo1 (blot 5) or MAb clo6 (blot 7) vs. the amounts of protein loaded.

[4] Determine the linear range of the integral vs. protein plots from [3] and pick one integral value each in this linear range which you shall use to determine the absolute amounts (μg) of CYP4A1 and CYP4A2/4A3 protein from the corresponding calibration curves [2]. Finally express the absolute amounts of CYP4A1 and CYP4A2/A3 as μg/mg microsomal protein and nmol/mg microsomal protein, respectively.

Reference

1. Laemmli UK (1970) Cleavage of structural proteins during the assembly of the head of bacteriophage T4. Nature 227:680-685

Acknowledgements

The authors wish to thank Ms. Ch. Anderhub for typing the manuscript and Bio-Rad Laboratories AG, Glattbrugg, Switzerland, for the permission to incorporate parts of their "Mini-Protean II Dual Slab Cell Instruction Manual" in this manuscript.

9 Measurement of the in Vitro Glucuronidation of Peroxisome Proliferator Carboxylic Acids by Liver Microsomes and Genetically Modified V 79 Cell Line

H. GOUDONNET, J. MAGDALOU and S. FOURNEL-GIGLEUX

9.1 Introduction and Aims

Peroxisome proliferators (PP) belonging to the series of carboxylic acids are hypolipidemic drugs (clofibric acid, ciprofibrate, fenofibrate,.....) or plastifiers (2-ethylhexanoic acid). In man, these substances are essentially excreted as acylglucuronides, which are formed by uridine-5'-diphosphate-(UDP)-glucuronosyltransferases (UGTs, EC 2.4.1.17), according to the following reaction:

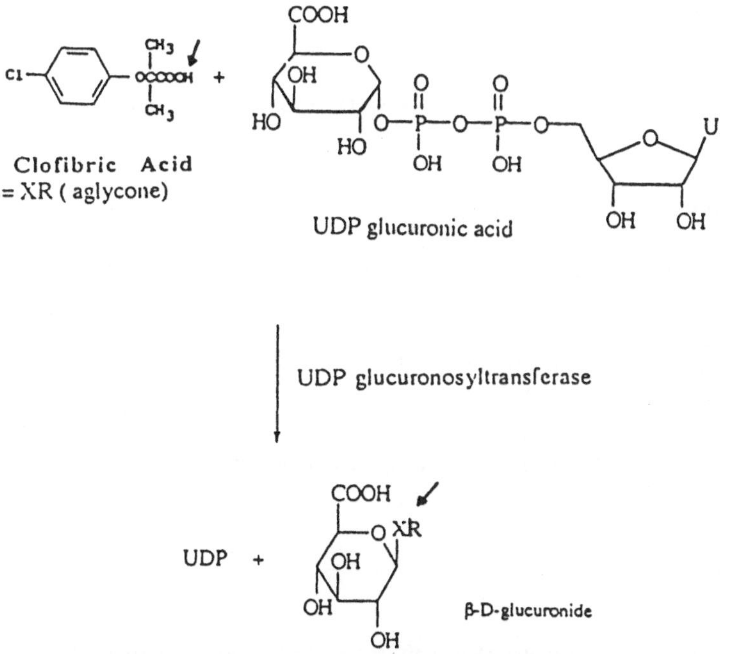

By contrast to ether glucuronides, ester (acyl) glucuronides are reactive metabolite species. At physiological pHs, they can undergo several non desirable reactions such as isomerization, hydrolysis, and transacylation, which can lead to the formation of covalently bound adducts with proteins [5]. This has been proposed as the mechanism which causes immunological reaction in patients receiving certain carboxylic acid drugs, other than fibrates, which has led to such drugs being withdrawn from the market. They are therefore toxicologically relevant.

In order to better understand the formation mechanisms of these glucuronides and to determine the UGTs isoforms that are involved in the process, it becomes necessary to develop in vitro metabolic systems. For that purpose, hepatic microsomes from various animal species and from man, as well as "transgenic" cells expressing a cDNA encoding a given isozyme, are two models which are widely used in the development of a new drug.

The aims of the experiment are:

- to follow and measure the glucuronidation reaction with carboxylic acids and PP as substrates; to optimize the enzymatic reaction;
- to determine the stereoselectivity of the glucuronidation, when the substrate has a chiral center;
- to identify, separate and quantitate glucuronides and isomers by analytical methods, such as thin layer chromatography (TLC) and high performance liquid chromatography (HPLC).

9.2　Equipment, Biological Sources, Chemicals and Solutions

9.2.1　Equipment

9.2.1.1 Equipment for Incubation

- a microcentrifuge for Eppendorf tubes
- a thermostated water-bath with agitation (37°C)
- a vortex mixer
- 1,5 ml Eppendorf tubes and racks
- adjustable automatic pipettes (Gilson or Labsystems) and tips
- ice and a container

9.2.1.2 Equipment for TLC

- a scanning system for β-radiometry (Bioscan Imaging Scanner, Packard)
- 20 x 20 cm silica gel plates
- a chromatography chamber for TLC
- 20 x 25 cm films for autoradiography (Kodak XAR-5)
- 20 x 25 cm exposure cassettes
- a freezer (-80°C)

- a device to blow TLC plates under compressed N_2 or air
- Hamilton syringes

9.2.1.3 Equipment for HPLC

- a 4 x 250 mm C_{18} columm packed with 7 μm Lichrosorb RP-18 particles and equipped with a pre-columm (Merck)
- an HPLC pump (Kontron model 325)
- a flow radiochromatograph detector (Flow On model A500, Packard)
- an HPLC UV detector (Kontron model 332)
- a control system and a dual channel integrator (Data System 450 MT, Kontron)
- Hamilton syringes

9.2.2 Biological Sources

9.2.2.1 Microsomes

Male Sprague Dawley rats (180-200g) were from Iffa Credo (St. Germain l'Abresle, France). The rats were treated either by intraperitoneal injection of phenobarbital (a specific UGT2 B1 isoform inductor) dissolved in saline (80mg/kg), for 5 days, or by 3-methylcholanthrene (a specific UGT1A1 isoform inductor) dissolved in corn oil (80 mg/kg; one per os administration). After 5 days, the rats were killed by decapitation, and liver microsomes were prepared by differential ultracentrifugation (100,000g fraction) and stored at -80° in 75 mM Tris-HCl, pH 7.4, until used.

The protein concentrations of liver microsomes (x mg/ml) were measured using the Lowry method.

9.2.2.2 V79 Cell Line Homogenates

V79 chinese hamster lung fibroblasts were transfected with a recombinant plasmid carrying the UDPGTr-2 cDNA , which encodes a phenobarbital-inducible rat UGT (UGT-2B1 isoform), expressed predominantly in the liver [4], and a vector including a neomycin resistance gene [3]. Some thirty resistant colonies were isolated after culture in selection medium, and these were screened for UGT activity towards 4-methylumbelliferone (4MU), a known substrate for UGT-2B1 [2].

The substrate specificity of UGT-2B1 was determined in homogenates of V79 UDPGTr-2 cells expressing relatively high levels of the enzyme (specific activity with 4 MU as substrate of 2.06 ±0.23 nmoles glucuronide formed per min per mg protein, n=3) using the general assay method of Bansal and Gessner [1]. The range of substrates glucuronidated was found to include testosterone, chloramphenicol and 4-hydroxybiphenyl.

In addition, UGT2B1 was able to catalyse the glucuronidation of a wide variety of carboxylic acids, including profen (NSAIDs), fibrate hypolipidemic agents, lithocholic acid, and members of the homologous series $CH_3(CH_2)_nCOOH$. A marked stereoselectivity was observed in the formation of profen acylglucuronides.

In the V79 cell line, cells were harvested and homogenized by freezing/thawing cycles, and proteins were measured (x mg/ml).

9.2.3 Chemicals

- Triton X100
- Tris
- $MgCl_2$
- dimethylsulfoxide (DMSO)
- saccharonolactone
- UDP-glucuronic acid (Boehringer)
- [U-^{14}C] UDP-glucuronic acid (180mCi/mmol, NEN Research Products, Dupont, 10 µCi)
- butyl-PBD (Packard)
- acetonitrile for HPLC (Carlo Erba RS)
- scintillation solution (Pico-Fluor 15, Packard)
- ethanol, n-butanol, acetic acid, acetone, 30 % aqueous ammonia, trifluoroacetic acid (analytical reagent purity)

9.2.4 Solutions

- 100mM Tris-HCl buffer, pH 7.0, containing 20 mM $MgCl_2$ and 10mM saccharonolactone (inhibitor of ß-glucuronidase) (100ml)
- 10 mg/ml Triton X100 (20ml)
- 0.07 mM [U-^{14}C] UDP-glucuronic acid
- 1 mM UDP-glucuronic acid (10ml)
- 20mM UDP-glucuronic acid (10ml)
- 10mM carboxylic acid: clofibric acid or ciprofibrate or 2-ethylhexanoic acid (racemate, R- and S+ forms, dissolved in DMSO (2ml)
- 6N HCl (100ml)

9.3 Experimental Procedure

9.3.1 Procedure for TLC

9.3.1.1 Incubation

- Optimal activation of microsomal UGT by Triton X-100: dilute microsomes in 100 mM Tris-HCl buffer to 5 mg/ml and add Triton X-100 (10mg/ml) or digitonine (100 mg/ml) at volume ratios of 1:1. Vortex the mixture and keep on ice for 20 min.
- For one assay, add the following sucessively in an Eppendorf tube (100 μl, final volume of the incubation) on ice:

  - 50 μl 100mM Tris buffer, pH 7.0
  - 5 μl 10 mM carboxylic acid solution
  - 20 μl activated microsomes (about 50 μg protein) or V79 cell homogenate (200 μg protein)

  Run a blank with labelled UDP-glucuronic acid without microsomes simultaneously.

- After mixing, leave in a water bath for 2 min at 37°C (temperature equilibration). Add 15 μl of 1mM UDP-glucuronic acid and 10 μl of labelled UDPGA (250,000 cpm). Vortex. Incubate for 30 min at 37°C.
- Add 50 μl ethanol (- 20°C) to stop the reaction. Vortex. Centrifuge for 5 min at 7,500 rpm.

9.3.1.2 TLC and Measurement of Radioactivity

- Put 60 μl of supernatant on the concentration zone of the silica gel plate. Dry. Develop in the solvent system n-butanol/acetic acid/acetone/30 % aqueous ammonia/water 70:78:50:1.5:60 (v/v). Migration time : about 4 h.
- After migration, dry the plate and pulverize it with butyl-PBD solution as enhancer. Place the film in contact with (onto) the plate and leave for 2-3 days in the dark. Develop and dry the film (see 4.6 - Autoradiography).
- As an alternative method, the plate can be scanned with a scanning system for β-radiometry.
- By superposition, and using the autoradiogram, mark the radioactive spots with a pencil. Scrap the silica gel carefully after adding a drop of water. Put the gel in liquid scintillation vials with 6 ml scintillation solution, and count in a liquid scintillation spectrometer.

9.3.2 Procedure for HPLC

9.3.2.1 Incubation

- For one assay, add the following in an Eppendorf tube on ice:

  - 250 µl Tris buffer pH 7.0,
  - 100 µl activated (see 9.3.1.1) microsomes (500 µg - 1 mg protein) from rat liver after treatment with phenobarbital, or V79 homogenate (500 µg protein),
  - 40 µl 10 mM clofibric acid in DMSO.

    Run a blank simultaneously.

- Mix and add 100 µl 20 mM UDP-glucuronic acid and 10 µl labelled UDP-glucuronic acid (250,000 cpm). Incubate for 20min at 37°C.
- Stop the reaction by 50 µl 6N HCl in ice and mixing. Centrifuge for 3 min at 7,500 rpm.

9.3.2.2 Reversed-Phase HPLC and Measurement of Radioactivity

- Inject an aliquot (50-100 µl) of the supernatant into the HPLC apparatus. The mixture acetonitrile-trifluoroacetic acid-water 55:0.08:145 (v/v/v) is used as the mobile phase at a continuous flow of 1.0 ml/min.
- Monitor the separated products using the UV detector at 233 nm for conjugated clofibric acid, and measure radioactivity using the flow radiochromatography detector.

9.3.3 Quantification of the Glucuronides

Calculate the quantity of glucuronide synthetised from the specific activity of the labeled conjugate.

References

1. Bansal SK, Gessner T (1980) A unified method for the assay of uridinediphosphoglucuronosyltransferase activity towards various aglycones using uridinediphospho(U 14C) glucuronic acid. Anal Biochem 109:321-329
2. Burchell B, Nebert DW, Nelson DR, Bock KW, Iyanagi T, Jansen PLM, Lancet D, Mulder GJ, Chowdhury JR, Siest G, Tephly TR, Mackenzie PI (1991) The UDP-

glucuronosyltransferase gene superfamily: suggested nomenclature based on evolutionary divergence. DNA Cell Biol 10:487-494

3. Mac-Fournel-Gigleux S, Sutherland L, Sabolovic N, Burchell B, Siest G (1991) Stable expression of two human UDP-glucuronosyltransferase cDNAs in V79 cell cultures. Mol Pharmacol 39:177-183

4. Mackenzie PI (1987) Rat liver UDP-glucuronosyltransferase: identification of cDNA encoding two enzymes with glucuronidate testosterone, dihydrotestosterone and B estradiol. J Biol Chem. 262:9744-9749

5. Smith PC, MacDonagh AT, Benet JZ (1990) Covalent binding of Zomepirac glucuronide to proteins: evidence for schiff base mechanism. Drug Metab Dispo 18:639-644

10 Molecular Modeling and Drug Design: Application to Some Peroxisome Proliferator Agents

A. BENMBAREK

10.1 Introduction

Every chemist has used CPK or Dreiding models to represent the structure of a molecule. Even if this representation gives an approximated idea on how the molecule is arranged in space, we have no idea on how this representation is stable in terms of intra-molecular interactions. In fact, many properties cannot be represented by such tools, most especially electronic parameters: charges, hydrophobicity, electronic conjugation, etc....

With the coming of age of computers [Silicon Graphics, Hewlett Packard or IBM RS600 workstations] and molecular modeling software [MAD/TSAR, Insight/Discover, Sybyl or Quanta/Charmm], a new level has been reached concerning the representation of molecules and their properties. Nevertheless, molecular modeling software is still developed on the basis of approximations. We can cite the approximations made on the Schrödinger equation for the description of an electronic system of a molecule. This is the reason why it is necessary to always keep in mind that, even though computers can takle very complicated systems, and perform sophisticated calculations, conclusions are always susceptible to questioning. Computer-Assisted Molecular Modeling is only a tool to reduce the part of empiricism. It is not a panacea.

10.2 Methodologies and Aims

The discovery of new drugs with desired biological activity can be done using three different approaches:

10.2.1 General Screening

The aim of this approach is the use of a large amount of different molecules in a biological trial. Then all the synthesized molecules are tested, even if they are not designed in the context of the concerned biological activity. Once a desired compound is found, a structure-activity program is initiated in order to seek the important functions for the concerned biological activity. When these essential elements are found, an optimization procedure is started. It consists of the

159

introduction of modifications, functional and structural, in order to enhance a better selectivity and potency. Angiotensin II antagonists domain is an example of this approach (for a review see [4]).

The functional modifications can be made in terms of isosterism. It is known that a carboxylate moiety can be replaced by a tetrazole or a sulfone fragment, a ter-butyl moity can be replaced by another hydrophobic fragment like cyclohexane, or an aromatic phenyl can be replaced by a thiophen one. Generally, this is accompanied by an evolution of the affinity and/or the selectivity, depending on the actual conditions of the interaction of the drug with its receptor.

The structural modifications concern only the spacers, without touching the essential functions. Those spacers are replaced by others. The goal is to introduce or supress rigidity in the structure. This supression or introduction of inflexibility is performed in order to make the products more active. Ideally, it is hoped to uncover a constrained and active compound, which would give indications on the active conformation.

10.2.2 Active Analogue Approach

The basis of this approach is the common idea of "lock and key". This methodology necessitates the knowledge of some active and inactive compounds which are closely related in term of structure. The problem then is to find out the common "pharmacophore ". This represents a set of chemical functions associated with an exact spatial position for each of them. The pharmacophore would represent the precise conformation of the ligand-receptor complex [7].

The above objective is generally achieved by the comparison of the conformational space of each molecule and the finding out of the common features concerning the subset of active molecules, on the one hand, and the conformational differences between the subsets of active and inactive molecules, on the other.

10.2.3 Direct Drug Design

It is necessary to have information concerning the exact three-dimensional structure of the active site of a receptor or an enzyme. The experimental data needed to describe three-dimensional structures of receptors are increasingly available. They are generally deposited in the Brookhaven Protein Data Bank, which now contains more than one thousand 3-D structures. One can obtain those structures free of charge from the Chemistry Department, Building 555, Brookhaven National Laboratory, Upton, NY 11973 USA; Phone number: 516-282-3629; Facsimile: 516-282-5751; E-Mail: pdb@bnlchm.bitnet or pdb@chm.chm.bnl.gov. It is also possible to contact the affiliated center in Germany: EMBL, Heidelberg, Germany; Phone: 49-6221-387-247; E-Mail: peter.rice@embl-heidelberg.de.

Using homology computer software, it is also possible to build a three-dimensional model of a receptor which has its primary sequence known. The aim of this approach is to build the model using a homology procedure applied to a close family of receptors for which one or more 3-D structures are known [9].

Once a hypothetical or established active site is defined, one can use molecular modeling tools to visualize the properties of this active site: chemical functions, their disposition and the space available for ligands. The modeler then has to try to fill the active site by complementary functions and connect them by bridge chemical fragments [2].

Even if this approach looks powerful, one cannot immediately certify the finding of an active compound because of other considerations such as bioavailability, dynamic aspects, allosteric sites, etc....

This last is an important point on which we will constantly focus during this drug design course. Computer-Assisted Drug Design gives us partial information. The essential part of such information's contribution is the reduction of the time necessary for the convergence to active compounds. There rest many limitations:

- influence of the surrounding medium (solvent, vacuum, membranes, receptor, etc...),
- metabolism problems,
- ability of the systems to represent the physico-chemical and biological properties, etc...

This course will be devoted to the application of some of the above techniques in studying a set of peroxisomal proliferator agents. The aim of this course is to give students who are not familiarized with molecular modeling methodologies an insight into how these could be helpful for the drug discovery process and how to use them for their own research projects.

10.3 Biological Molecules

The studied molecules are methyl- and ethyl-hexanoic acids (Figure 1). They are metabolites of plasticizers, which can be found in food. It is thus necessary to characterize such molecules in terms of toxicity before putting plasticizers on the market. Trials are done on animals and human cell cultures. The results in terms of relative toxicity are presented in Figure 1. The reference value, 100, represents the compound 2EHA. In this practical, an attempt will be made to correlate the observed toxicity of the compounds in question to their structural properties. This may prove helpful for:

- the estimation of potentialities of new plasticizers, such as resistance to heat,
- the design of new plasticizers with less toxicity, and
- knowledge gains about peroxisome proliferators.

Molecule	Structure	Relative Toxicity
2EHA	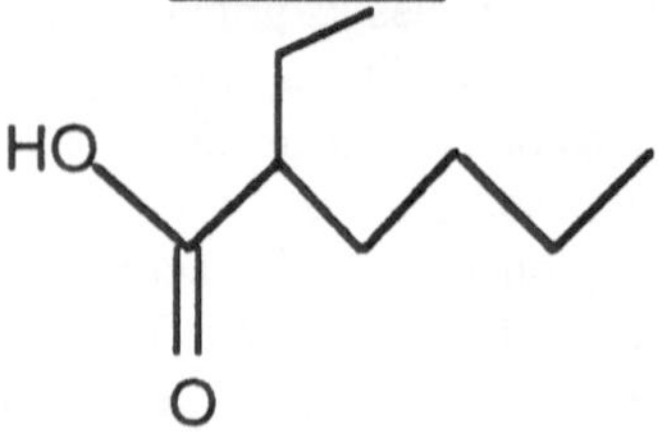	100
3EHA	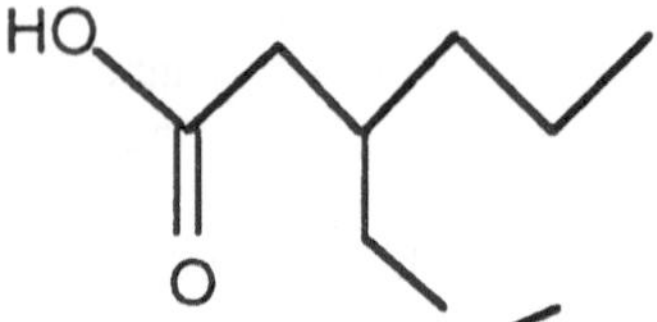	56
4EHA	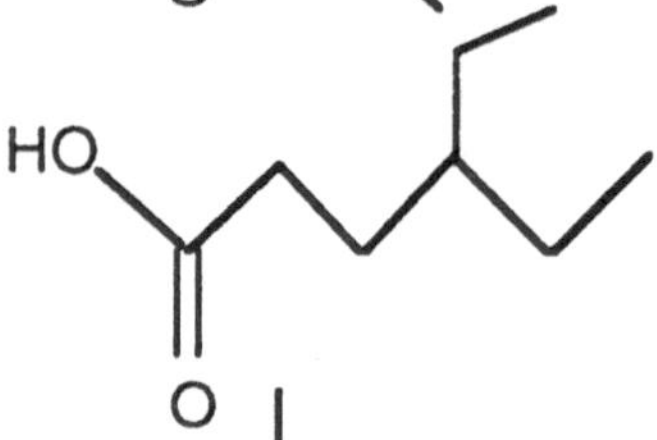	60
2MHA	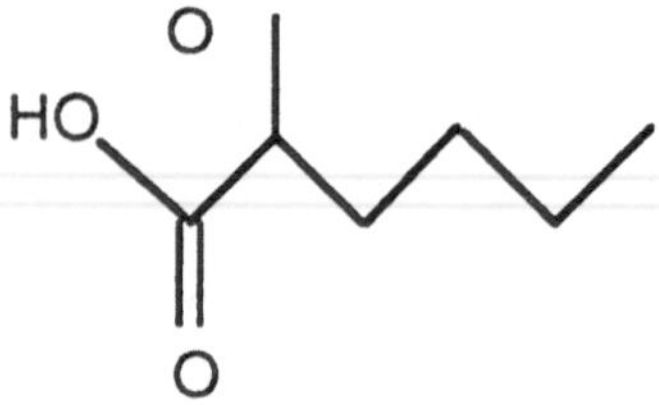	52
3MHA	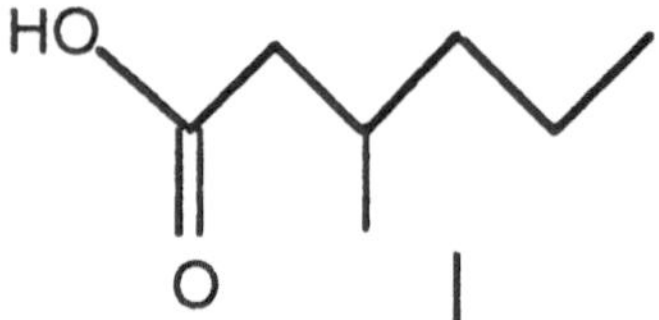	34
4MHA	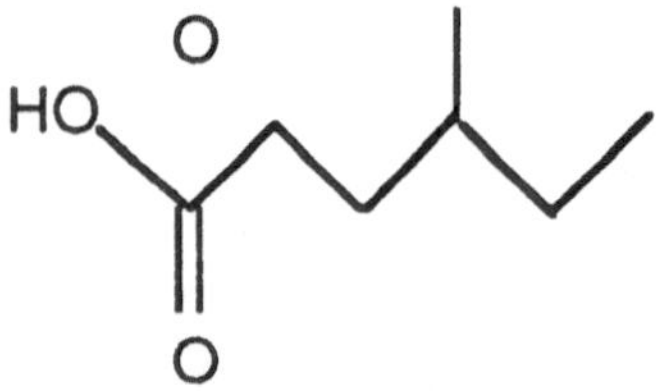	13
5MHA	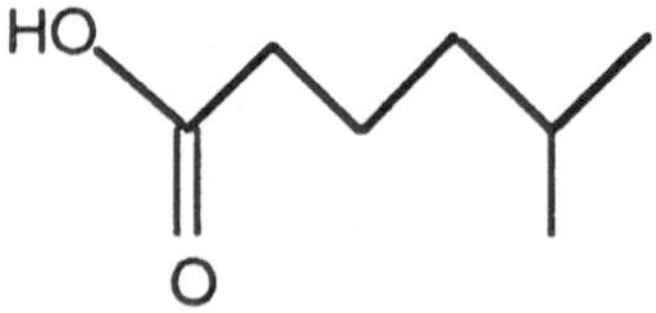	9

Figure 1: Structures of methyl-hexanoic (MHA) and ethyl-hexanoic (EHA) acids and their relative toxicity.

10.4 Equipment and Procedures

The work will be accomplished on an IBM Risc-6000 workstation, and the software used for the calculations will be MAD (*M*olecular *A*dvanced *D*esign) of Oxford Molecular, Ltd.

The course will be divided into two parts:

- A fundamental course with discussions (1 h).
- A practical session (2 h). This will include:

  - building of molecular structures,
  - minimizing of energy,
  - calculation of molecules' properties: lipophilicity, charges, volume, etc...,
  - fitting of structures (volume differences),
  - sampling of conformational space,
  - looking for the most stable conformers, and
  - attempting to relate combinations of properties to biological activities of the structures.

10.5 Results and Conclusions

All of the structures were built and fully optimized in terms of energy, and their conformational space was sampled using a random conformational search procedure. The conformers associated to the lowest energy value were stored for each molecule.

All of the molecules present a common carboxylate moiety, so they were superimposed on the basis of the common COO fragment. The visualization of the superimposed structures with their isopotential electrostatic field showed the existence of two pockets: a hydrophobic pocket and a hydrophylic pocket. Furthermore, the shape of the iso-contour surface corresponding to the alkyl chain differed from one structure to the next, while the iso-contour surface of the acidic moiety remained the same for all the structures.

The discussions were then focused on the importance and possible role of the hydrophobic alkyl chain. Figure 2 shows the clipped volumes of the reference structure 2EHA and the two structures, 2MHA and 5MHA, presenting 52% and 9%, respectively, as relative toxicity.

It is easy to observe that the decrease of the relative toxicity corresponds to the decrease of the length of the sub-chain on position 2. The discussions on this observation resulted in two questions:

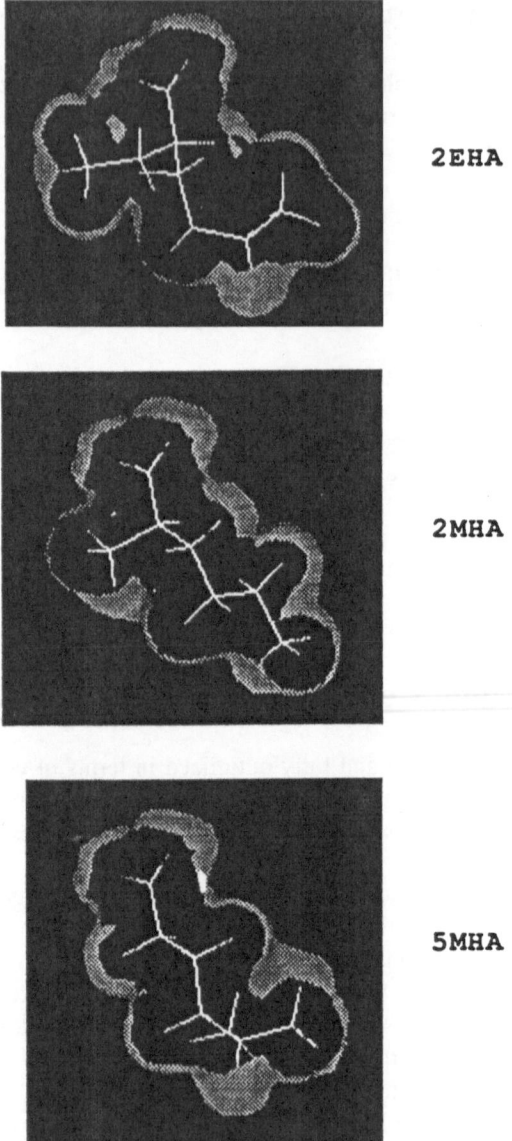

Figure 2: Clipped volumes of the reference compound 2EHA and the two compounds 2MHA and 5MHA.

- Does the alkyl chain play a protective role for interactions of the acid group with a positive pole (Figures 3 and 4)?
- Is the shape of the hydrophobic alkyl chain important, so that the alkyl sub-chain on position 2 must reach a second hydrophobic pole (Figure 5)?

Concerning the first question, it was found in the litterature [3] that the crystallographic structure of the phosphocholine-Fab Mc PC603 complex presents an ion pair surrounded by a hydrophobic cluster, leading thusly to a more stable complex. Other references [5,6,7] concerning such stabilizing systems can also be reviewed. In our case the carboxylate moiety would be protected or not protected, depending on its lateral accessibility.

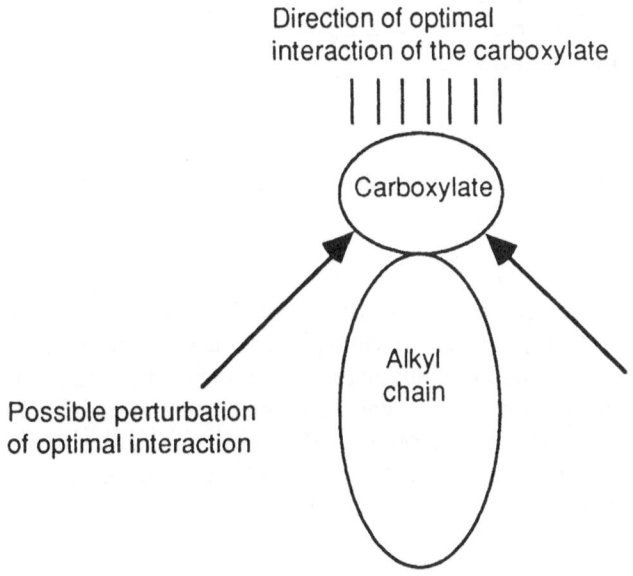

Figure 3: The interaction of the carboxylate moiety can be perturbed laterally by a positively charged fragment(s).

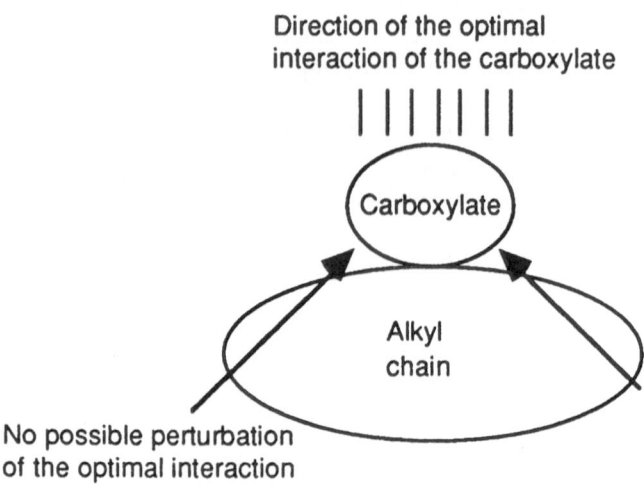

Direction of the optimal
interaction of the carboxylate

Carboxylate

Alkyl
chain

No possible perturbation
of the optimal interaction

Figure 4: The interaction of the carboxylate moiety cannot be perturbed laterally by a positively charged fragment(s).

Concerning the second question, a maximum of interaction of the whole molecule would be reached by a system of three points: one point of electrostatic nature and the two others of hydrophobic nature (Figure 5). The results of the structure-activity presented in this study suggest that the branching must occur on the position number 2. The conclusion was that rigid molecules must be synthesized in order to verify the previous hypothesis. One can imagine structures such as the following ones:

166

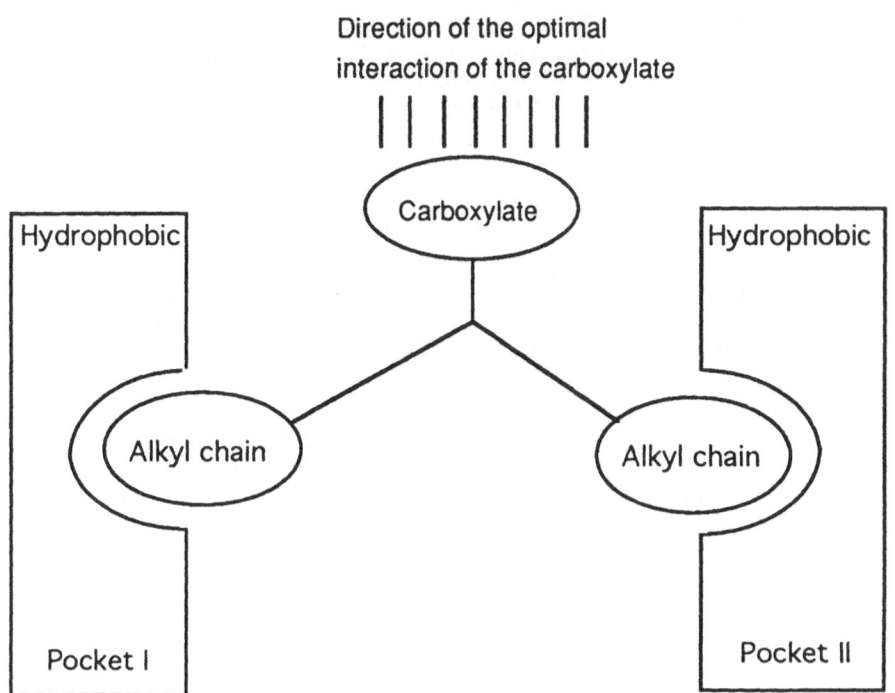

Figure 5: The second hypothetical model of interaction: the whole molecule is stabilized by three poles, one of electrostatic nature and the two others of hydrophobic nature.

Acknowledgements

The author wishes to thank Prof. J.C. Lhuguenot (ENSBANA, Dijon) for providing the results related to toxicology used in this study. A special thanks as well to his collaborator Chrystel Varnier for her assistance with the practical sessions.

References

1. Bohm MJ (1991) The computer program LUDI: A new method for the de novo design of enzyme inhibitors. J Comp Aided-Molecular Design 6:61-78
2. Desjarlais RL, Sheridan RP, Seibel GL, Dixon JS, Kuntz ID, Venkataraghavan R (1988) Using shape complementarity as an initial screen in designing ligands for a receptor binding site of known three-dimensional structure. J Med Chem 31:722-729

3. Dougherty DA, Stauffer D (1990) Acetylcholine binding by a synthetic receptor: implications for biological recognition. Science 250:1558-1560
4. Hodges JC, Hamby JM, Blankley JC (1992) Angiotensin II receptor binding inhibitors. Drugs of the Future 17:575-593
5. Mian S, Bradwell A, Olson A (1991) Structure, function and properties of antibody binding sites. J Mol Biol 217:133-151
6. Satow Y, Cohen G, Padlan E, Davies D (1986) Phosphocholine binding Fab McPC603. An x-ray diffraction study at 2.7Å. J Mol Biol 190:593-604
7. Seiler MP, Markstein R, Boelstereli JJ, Walkinshaw MD (1989) Characterisation of dopamine subtypes by comparative structure-activity relationships. Mol Pharmacol 35:643-651
8. Sussman JL, Harel M, Frolow F, Oefner O, Goldman A, Toker L, Silman J (1991) Atomic structure of acetylcholinesterase from Torpedo californica: a prototypic acetylcholine-binding protein. Science 253:872-879
9. Trumpp-Kallmeyer S, Hoflack J, Bruinvels A, Hibert M (1992) Modeling of G-protein coupled receptors: Application to dopamine, adrenaline, serotonin, acetylcholine and mammalian opsin receptors. J Med Chem 35:3448-3462

IV Cell Culture and Genetic Diseases

11 Fao Cell Line as a Model for the Study of the Effect of Peroxisome Proliferators on Cellular Functions

C. BROCARD, L.C. RAMIREZ and P. BOURNOT

11.1 Introduction

Peroxisomes have essential functions in lipid metabolism. They are involved in the β-oxidation of a variety of compounds, the synthesis of cholesterol and bile acids, the synthesis of ether lipids, and several other pathways (for review, see [9, 15, 8]). A characteristic of peroxisomes in rodents is their proliferative response and the induction of fatty acid ß-oxidation enzymes upon treatment with several structurally unrelated chemicals, such as hypolipidemic drugs and plasticizers [6, 7, 12, 10].

In vitro systems are very useful for the study of the mechanism of peroxisome proliferation and ß-oxidation induction. The H4IIEC3 cell line and the 7800 C1 Morris hepatoma cells have a stable peroxisomal acyl-CoA oxidase activity, and their ß-oxidation is inducible [11, 13]. Acyl-CoA oxidase activity has been detected in two other rat hepatoma cell lines, Fao and MH_1C_1, and in one human hepatoblastoma cell line, HepG2, and is induced by ciprofibrate, a hypolipidemic drug ([2] in press).

Here, we will study the response to ciprofibrate and bezafibrate of the Fao line, known to be well-differentiated [4, 5]. The peroxisomal matrix enzyme, acyl-CoA oxidase, which catalyses the first and rate-limiting step in the ß-oxidation pathway of fatty acids, is widely used as a marker of the cellular response to peroxisome proliferators. We will determine its activity in cells treated with the proliferators by a sensitive and specific fluorometric assay in which the H_2O_2 produced reacts with homovanillic acid in the presence of peroxidase to form a fluorescent dimer.

By contrast, the peroxisomal membrane enzyme dihydroxyacetone-phosphate acyltransferase (DHAP-AT), which catalyses the first step of ether lipid biosynthesis, is not specifically inducible by peroxisome proliferators [3]. The DHAP-AT activity will also be measured in the samples of control and treated cells, to serve as a comparison (see experiment 12).

11.2 Equipment, Chemicals and Solutions

11.2.1 Equipment

11.2.1.1 Access to

- an autoclave
- a freezer (-70°C)
- an incubator at 37°C, with a water-saturated, 5% CO_2/95% air atmosphere (Jouan series EB)
- a laminar flow hood (Flufrance, Supcris 15) with a Bunsen burner
- a low speed centrifuge (Jouan B 3.11)
- a phase-contrast microscope with a squared counting grid on the occular (Olympus CK 2 TR) equipped with a camera or a Sony video equipment
- films, black and white, 400 ASA, 12 poses
- automatic pipetting aid devices (Sarstedt) and supports
- adjustable automatic pipettes (Gilson or Labsystems) with sterile and non-sterile tips and a carrousel support
- a sonicating bath (Ney, Ultrasonik 300)
- a sonicator probe (Bioblock, Vibra cell 72434)
- a spectrofluorometer (Kontron SFM 25) connected to a thermostated water bath
- a visible spectrophotometer (Kontron, Uvikon series)
- ice and containers

11.2.2 Disposable and Other Material

11.2.2.1 Disposable Sterile Material

- 15 ml conical centrifuge tubes (Falcon 2097)
- 25 cm^2 culture flasks (Nunc 1 63371)
- 80 cm^2 culture flasks (Nunc 1 53732)
- 1 ml, 5 ml and 10 ml pipettes
- 1 ml syringues with needle

11.2.2.2 Other Material

- disposable microcuvettes for spectrophotometry
- 1.5 ml Eppendorf tubes
- 20 and 500 µl Hamilton glass syringes
- Mallassez counting slides
- a 250 ml metal beaker
- 2 ml glass tubes

172

- quartz fluorometric microcuvettes
- racks for 5 and 15 ml tubes

11.2.3 Chemicals

- bezafibrate (Oberval)
- ciprofibrate (Sterling-Winthrop)
- dimethylsulfoxide (DMSO; Sigma)
- fatty acid free bovine serum albumin (BSA; Boehringer)
- streptomycine (Diamant)
- specilline G (Specia)
- trypan blue (Sigma)
- 30 % hydrogen peroxide, pure, stabilized (Merck)
- peroxidase (HRP type II, Sigma, 25,000 U)
- flavin adenine dinucleotide (FAD; Boehringer, 10 mg)
- palmitoyl-CoA (Pharmacia, 25 mg)
- homovanillic acid (Aldrich, 100 mg)
- glycylglycine (Boehringer)
- sodium chloride
- potassium phosphate
- sodium bicarbonate
- sodium azide

11.2.4 Culture Medium and Solutions

11.2.4.1 Cell Culture

- 1 M bezafibrate (72 mg/200 µl)
- 1 M ciprofibrate (58 mg/200 µl)
- fetal calf serum (FCS; Gibco BRL)
- sterile Ham F12 medium containing 146 mg/L of glutamine (Gibco BRL), supplemented with 1.25 U/ml specilline G and 0.125 mg/ml streptomycine
- MOPS buffer: 0.25 M sucrose, 1mM EDTA, 0.1% ethanol, 5mM MOPS, pH 7.4
- saline solution (HBSS, Gibco BRL)
- 0.4% trypan blue, prepared in 0.81% sodium chloride and 0.06% potassium phosphate
- trypsin solution containing 0.5 g/L trypsin and 0.2 g/L EDTA (Gibco BRL 0430 5300M, 500ml)

11.2.4.2 Protein Measurement

- a Bio-Rad protein assay kit
- 0.1 mg/ml of fatty acid free BSA

11.2.4.3 Acyl-CoA Oxidase Activity Determination

- 1 mM fatty acid free BSA (65 mg/ml)
- 1mM FAD (8.3 mg/10 ml)
- 0.5 mM FAD (4.15 mg/10 ml)
- 0.5 M glycylglycine buffer, pH 8.3
- 27.5 mM homovanillic acid (5 mg/ml)
- 5 mM palmitoyl-CoA (5.03 mg/ml)
- 1 mg/ml peroxidase in glycylglycine buffer
- 10 mM sodium azide (0.65 mg/ml)

11.3 Experimental Procedures

The Fao cell line, derived from Reuber H35 rat hepatoma [4], was obtained from Dr. J. Deschatrette. Cells are grown in Ham F12 medium supplemented with 5% FCS. All media are supplemented with 1.25 U/ml specilline G and 0.125 mg/ml streptomycine. Cells are seeded and grown in 25 cm^2 plastic flasks and placed in a 37°C incubator with a water saturated atmosphere equilibrated with 5% CO_2/air. The culture medium is changed every day during 4 to 5 days, and the experiments are performed during the exponential growth period of the cells.

Day 1 (2 h)

11.3.1 Observation, Upkeep and Treatment of Cells

Two series of experiments, A and B, are to be carried out in parallel, the cells being treated with either ciprofibrate or bezafibrate at 2 concentrations:

- Experiment A: 1 cell flask as control, 2 cell flasks treated with ciprofibrate at concentrations of 100 and 250 μM and 1 cell flask reserved for trypsination and seeding new flasks.
- Experiment B: same as in A, except that the cells are treated with the proliferator bezafibrate.

- Prepare four different types of medium:

  - Control medium: 20 ml of 5% FCS supplemented Ham F12 containing 0.1% DMSO.
  - Treatment media:
    - 20 ml of 5% FCS supplemented medium containing 100 μM proliferator and 0.1% DMSO.
    - 20 ml of 5% FCS supplemented medium containing 250 μM proliferator and 0.1% DMSO.
  - Upkeep medium: 40 ml of 5% FCS supplemented medium.

- Prepare the control and treatment media by adding the appropriate volumes of a stock solution of proliferator at 1 M to Ham F12 medium containing the antibiotics and then adding DMSO to reach a final solution of 0.1% DMSO. Place the three media in a sonicating bath for 10 min at room temperature before adding 5% FCS. Prepare the upkeep medium directly without sonication.
- Remove the cell flasks from the incubator oven and close their lids tightly. Observe the aspect and degree of confluence of the different cell flasks under a phase-contrast microscope and photograph a representative area of each cell flask.
- Replace the medium of each flask by 4 ml of the appropriate medium, and place the slightly opened flasks in the incubator.

Day 2 (30 min)

Remove the cell flasks from the incubator oven and close their lids tightly. Observe the cells under the microscope. Replace the spent medium of each flask with 4 ml of the appropriate medium and place the slightly opened flasks in the incubator.

Day 3 (30 min)

Observe the cells and change the media, as described for *Day 2*.

Day 4 (2-3 h)

11.3.2 Cell Harvest After Treatment

- Remove the cell flasks from the incubator oven and close their lids tightly. Observe the cells under the microscope and photograph representative areas.

- Harvest the control and treated cells in order to determine the amount of protein and palmitoyl-CoA oxidase activity in the following manner: after removing the medium, wash the cellular monolayer three times with 4 ml HBSS solution, in order to remove all trace of serum, which might inhibit the activity of trypsin. Detach the cells in 2 ml of trypsin solution during 5 min at 37°C. Stop the reaction by addition of 2 ml of 5% FCS-supplemented Ham F12 medium (or 0.5 ml of serum) per flask.
- Transfer the cell suspensions into conical tubes; rinse the flask with 2 ml of medium, which is added to the cell suspension. Centrifuge the suspension for 10 min at 80 x g. Resuspend the cells in 4 ml MOPS buffer to wash and recentrifuge for 10 min at 80 x g (set aside an aliquot of 50 μl in a 2 ml tube for counting the cells). Finally, resuspend the cells in 500 μl MOPS buffer.
- Sonicate with a sonicator probe of 8 watts for 3 x 10 s in ice.
- Supplement the homogenates with FAD to a final concentration of 10 μM after sonication, using the 1 mM stock solution. Separate the samples into 3 aliquots and freeze them at -70°C.
- In order to count the cells, mix 50 μl of the cell suspension with 50 μl of a solution of trypan blue. Place this mixture on a Malassez counting slide, and count the cells. Calculate the total number of cells.

11.3.3 Trypsination and Seeding

- Calculate the total number of cells in the monolayer using the squared counting grid on the microscope.
- Wash and trypsinate the cells, as indicated above for the cell harvest.
- Centrifuge the cell suspension containing trypsin, EDTA and medium for 10 min at 80 x g. Resuspend the cells in 2 ml of serum-supplemented medium and count them on a Malassez counting slide by mixing 50 μl of the cell suspension with 50 μl of a solution of trypan blue.
- After calculating the number of live cells, those which exclude the trypan blue, seed approximately 5×10^5 cells per flask. Complete each flask to 4 ml with serum-supplemented medium and place the slightly opened flasks in the incubator.

Day 5 (6h)

11.3.4 Protein Assay

Measure the protein content in the cell homogenate according to the method of Bradford (1976) using the Bio-Rad protein assay:

- Prepare a blank, a BSA standard curve (0, 1, 3, 5, 8, 10, 16, 20, 30 µg) and 3 tests per sample with 5 µl cell homogenate, directly in disposable microcuvettes.
- Complete each cuvette to 0.8 ml with water and add 0.2 ml of reagent per tube. Mix well and let stand at room temperature for 10 min.
- Read the OD at 595 nm. Calculate the amount of protein in each sample.

11.3.5 Fluorometric Assay of Peroxisomal Palmitoyl-CoA Oxidase Activity

Assay palmitoyl-CoA oxidase activity by fluorometric measurement of the H_2O_2 produced according to the method previously reported [12]. The medium contains 50 mM glycylglycine buffer pH 8.3, 0.55 mM homovanillic acid (0.1 mg/ml), 0.02 mg/ml peroxidase, 3 µM FAD, 20 µM fatty acid free BSA, 0.2 mM sodium azide and the substrate, 100 µM palmitoyl-CoA, for a final volume of 0.5 ml. The reaction is started by the addition of cell homogenate, containing 20 to 100 µg of protein (carry out at 37°C). A blank is also incubated without substrate.

- Mix well in a quartz fluorometric microcuvette:

  - 50 µl of 0.5 M glycyl glycine buffer, pH 8.3
  - 397 µl of ultrapure water
  - 10 µl of 1 mg/ml peroxidase
  - 10 µl of 27.5 mM homovanillic acid
  - 10 µl of 1 mM free fatty acid BSA
  - 3 µl of 0.5 mM FAD
  - 10 µl of 5 mM palmitoyl-CoA
  - 10 µl of 10 mM sodium azide

- Incubate the microcuvette for 1.5 to 2 min at 37°C in the spectrofluorometer, add 10 µl of the cell homogenate and mix the contents of the cuvette by hand.
- Record the spectrometric data during the first 8 min. Use the following conditions for the fluorometric measurement : time drive, 8.00 min; time scale, 2.00 min/cm; excitation, 326 nm; emission, 425 nm; calibration, 100.0; high voltage, 400 V; blank, 5.00; factor, 4.0 or 8.0; response, 8.0 s.
- Convert the initial speed of the reaction into nmoles of H_2O_2/min after calibration with reference solutions of 5×10^{-5} M H_2O_2.

Day 6 (30 min)

11.3.6 Observation of Seeded Cells and Counting

Count the attached cells using the squared counting grid on the microscope and calculate the seeding efficiency.

11.4 Comments

The following results are expected from the treatment of the cells with the proliferators:

- a change in the size and aspect of the cells,
- a dose-dependent stimulation by the proliferator of the palmitoyl-CoA oxidase activity,
- a degree of stimulation dependent on the proliferator, and
- a possible change in the number of cells as a response to the treatment.

References

1. Bradford M (1976) A rapid and sensitive method for the quantitation of microgram quantities of protein utilizing the principle of protein dye binding. Anal Biochem 72: 248-254
2. Brocard C, Es-souni M, Ramirez LC, Latruffe N and Bournot P (1993) Stimulation of peroxisomal palmitoyl-CoA oxidase activity by ciprofibrate in hepatic cell lines: comparative studies in Fao, MH_1C_1 and HepG2 cells. Bio Cell 77:37-41
3. Causeret C, Bentejac M, Clemencet MC and Bugaut M (1993) Effects of two peroxisome proliferators (Ciprofibrate and Fenofibrate) on peroxisomal membrane proteins and dihydroxyacetone-phosphate acyl-transferase activity in rat liver. Cell Mol Biol 39:65-80
4. Deschatrette J, Weiss MC (1974) Characterisation of differentiated and dedifferenciated clones of rat hepatoma. Biochimie 56: 1603-1611
5. Deschatrette J, Moore EE, Dubois M & Weiss MC (1980) Dedifferenciated variants of a rat hepatoma: reversion analysis. Cell 19: 1043-1051
6. Hess R, Staubli W, Reiss W (1965) Nature of the hepatomegalic effect produced by ethyl-chlorophenoxy-isobutyrate in the rat. Nature 208: 856-859
7. Lazarow PB, de Duve C (1976) A fatty acyl-CoA oxidizing system in rat liver peroxisomes; enhancement by clofibrate, a hypolidemic drug. Proc Natl Acad Sci USA 73: 2043-2046
8. Mestdagh N, Vamecq J (1992) Mitochondrial and peroxisomal acyl-CoA β-oxidation in mammalian metabolism. In : Recent Advances in Cellular and

Molecular Biology (Wegmann RJ, Wegmann MA, eds) Peteers Press, Leuven - Belgium: 115-160

9. Osmundsen H,, Bremer J, Pedersen JI (1991) Metabolic aspects of peroxisomal ß-oxidation. Biochim Biophys Acta 1085: 141-158

10. Osumi T, Hashimoto T (1984) The inducible fatty acid oxidation system in mammalian peroxisomes. Trends Biochem Sci 9: 317-319

11. Osumi T, Yokota S, Hashimoto T (1990) Proliferation of peroxisomes and induction of peroxisomal ß-oxidation enzymes in Rat Hepatoma H4IIEC3 by ciprofibrate. J Biochem 108: 614-621

12. Reddy JK, Warren JR, Reddy MK, Lalwani ND (1982) Hepatic and renal effects of peroxisome proliferators: biological implications. Ann N Y Acad Sci 386: 81-110

13. Spydevold O, Bremer J (1989). Induction of peroxisomal ß-oxidation in 7800 C1 Morris hepatoma cells in steady state by fatty acids and fatty acid analogues. Biochim Biophys Acta 1003: 72-79.

14. Vamecq J (1990) Fluorimetric assay of peroxisomal Oxidases. Anal Biochem 186: 340-349

15. Van den Bosch H, Schutgens RBH, Wanders JA, Tager JM (1992) Biochemistry of peroxisomes. Ann Rev Biochem 61: 157-97

12 Measurement of Dihydroxyacetone-Phosphate Acyl-Transferase (DHAP-AT) Activity in Liver Peroxisomes and Cell Lines

C. CAUSERET, M. BENTÉJAC and M. BUGAUT

12.1 Introduction and Aims

The first three steps of ether-lipid biosynthesis are catalysed by enzymes located in the peroxisomal membrane, whereas the further steps are associated with the endoplasmic reticulum [4,9]. The peroxisomal enzymes are dihydroxyacetonephosphate (DHAP) acyltransferase, alkyl-DHAP synthase and alkyl-DHAP reductase.

Dihydroxyacetonephosphate acyltransferase (DHAP-AT, E.C. 2.3.1.42) is found only in peroxisomes [3] and, therefore, can be used as an enzymatic marker of this organelle [8]. DHAP-AT's active site is on the inner face of the membrane. The enzyme catalyses the first step in the synthesis of ether lipids, which have acyl-DHAP as an obligate precursor.

The role of peroxisomes in ether-lipid biosynthesis has been elucidated in the study of peroxisomal diseases, such as Zellweger syndrome, characterized at the subcellular level by the absence of normal peroxisomal structures and severe deficiency in ether-lipids. DHAP-AT activity is almost undetectable in the fibroblasts of Zellweger patients, and its measurement can be useful in the prenatal diagnostic of this disease.

In rodents (not in humans), peroxisome proliferators such as hypolipidemic drugs (clofibrate, bezafibrate, fenofibrate, ciprofibrate), are known to strongly increase the activities of some peroxisomal enzymes such as palmitoyl-CoA oxidase. In contrast, the specific activity of DHAP-AT is little affected by these proliferators [2, 6, 10].

DHAP-AT activity will be measured in liver homogenates and purified peroxisomes from control and bezafibrate-treated rats, hepatoma cells treated with proliferators (see protocol 11), and human fibroblasts from patients suffering different peroxisomal diseases (see protocols 13 and 14). Although the peroxisomal enzyme activities in hepatoma cells are lower than those in liver homogenates, hepatoma cell lines represent an interesting model to study peroxisomes [1, 11].

DHAP-AT activity is assessed by measuring the amount of acyl-[14C] DHAP formed from [14C] DHAP [5, 7]. Since [14C] DHAP is not available, the assay requires a preliminary step for its preparation from sn-[14C] glycerol-3-phosphate, as follows:

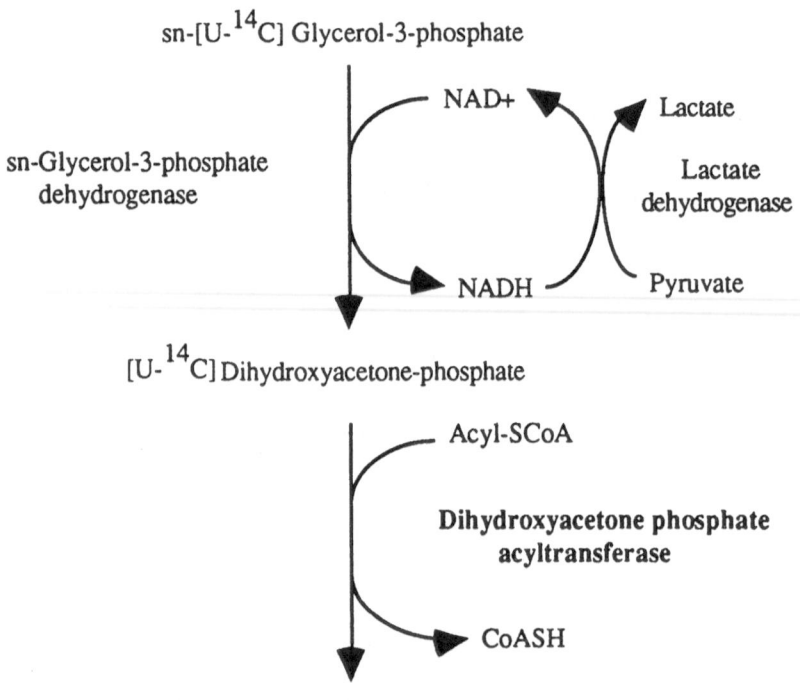

sn-[U-14C] Glycerol-3-phosphate

NAD+   Lactate

sn-Glycerol-3-phosphate dehydrogenase

Lactate dehydrogenase

NADH   Pyruvate

[U-14C] Dihydroxyacetone-phosphate

Acyl-SCoA

Dihydroxyacetone phosphate acyltransferase

CoASH

[U-14C] Acyl-dihydroxyacetone-phosphate

12.2 Equipment, Chemicals and Solutions

12.2.1 Equipment

- a scintillation counter (Beckman 6000 IC)
- a vortex mixer
- a low speed centrifuge (Beckman GPKR)
- a thermostated water bath for test tubes
- plastic trays
- glass test tubes (1.2 x 7.5 cm) and racks
- scintillation vials with glass microfiber filters (Whatmann GF/C)
- adjustable automatic pipettes (Gilson or Labsystems) and tips
- Hamilton syringes
- a 1 ml Hamilton syringe with a Teflon pestle
- a 25 or 50 µl adjustable automatic pipette (Microman M25 or M50, Gilson) plus tips with pestles
- Parafilm
- gloves
- Pasteur pipettes and bulbs
- a metal dustbin for radioactive waste
- a timer
- a container with ice
- a scintillation liquid dispenser
- scintillation vials

12.2.2 Chemicals

- L-lactate dehydrogenase (LDH; from rabbit muscle, 860 U/mg protein, EC 1.1.1.27, Sigma)
- sn-glycerol-3-phosphate dehydrogenase (G3P-DH; from rabbit muscle, 265 U/mg protein, EC 1.1.1.8, Sigma)
- sn-[U-^{14}C] glycerol-3-phosphate (159 mCi/mmol, Amersham)
- scintillation liquid (Pico-Fluor^{TM} 15, Packard)
- pyruvate
- sn-glycerol-3-phosphate (G-3P)
- free fatty acid bovine serum albumin (BSA; Boehringer)
- β-nicotinamide adenine dinucleotide (NAD; Sigma)
- palmitoyl coenzyme A (Pharmacia)
- Tris base
- sodium acetate
- magnesium chloride
- sodium fluoride

- trichloroacetic acid (TCA)
- chloroform
- methanol
- phosphoric acid
- potassium chloride
- ultra pure water

12.2.3 Solutions

- incubation medium n°1 containing 5 mM pyruvate, 1 mM β-NAD, 0.6 mM sn-G3P in 50 mM Tris-HCl, pH 7.6
- sn-[U-^{14}C] G3P diluted with sn-G3P to 16 nCi/nmol
- incubation medium n° 2 containing 8 mM NaF, 8 mM MgCl$_2$, 0.15 mM palmitoyl-CoA in 100 mM sodium acetate, pH 5.4, plus 0.4 mg BSA per assay
- incubation medium n° 2' as n° 2, but without palmitoyl CoA
- methanol/chloroform 2:1 (v/v)
- a solution containing 0.2 M H$_3$PO$_4$ and 2 M KCl
- 10%, 5% and 1% TCA solutions

12.3 Experimental Procedure

Each sample has first been assessed for protein using the Bio Rad protein assay kit. Three separate assessments and one control are necessary to determine DHAP-AT activity accurately in a sample.

Recommendation: radiochemicals have to be handled on a tray using gloves.

12.3.1 First Step: Synthesis of the Substrate (1.5 h)

- In a 5 ml glass test tube, add:

    - 25 µl incubation medium n°1 containing LDH (0.25 U/assay),
    - G3P-DH (0.25 U/assay), and
    - sn-[U-^{14}C]G3P (about 700 000 dpm/ assay).

- Close the tube with Parafilm, mix gently, and incubate at 25°C for 1 h. During the incubation period, prepare series of marked tubes for the second step.
- Stop the reaction by adding an equal volume of chloroform. Mix gently. Transfer the upper phase, which contains formed [^{14}C]DHAP, into another test tube using a Pasteur pipette. Close using Parafilm. [^{14}C]DHAP can be stored at +4°C for a whole day without degradation.

- Control the radioactivity level by adding 10 μl to a scintillation vial and 4.5 ml Pico-Fluor™ 15. Vortex and count.

12.3.2 Second Step: Enzymatic Reaction (1 h)

- In a test tube, add in the following order:

   - 95 μl incubation medium n°2 (3 tubes for a sample)
     or 95 μl incubation medium n°2' (1 control for a sample),
   - 10 μg protein (up to 50 μg if necessary), using a Hamilton syringe, and
   - 25 μl [^{14}C] DHAP, using a Hamilton syringe.

   Prepare one blank without protein (in a scintillation vial without filter).

- Close with Parafilm and mix gently.
- Incubate at 37°C for 30 min.

12.3.3 Extraction, Washing, and Measurement of the Labelled Product (2 h)

- After incubation, add the following into the test tubes, using a Hamilton syringe with a Teflon pestle:

   - 450 μl methanol/chloroform 2:1 (mix gently),
   - 150 μl chloroform (mix gently), and
   - 150 μl 2M KCl, 0.2M H_3PO_4 solution.

- Mix for 2 min.
- Centrifuge at 150 g for 5 min.
- (Discard the aqueous upper phase, but not the interface, with a Pasteur pipette.)
- Put down slowly 200 μl chloroform lower phase, which contains formed acyl-[^{14}C]DHAP, on a filter placed in the bottom of a scintillation vial, by four 50 μl-fractions, using a Microman pipette. Don't make a hole in the filter. After putting down each 50 μl-fraction on the filter, dry using an air-dryer or a stove. Note the volume of the remaining lower phase.
- Wash the well-dried filter in the scintillation vial, by adding 4 ml 10% TCA, and incubate for 30 min. Discard 10% TCA with a Pasteur pipette.
- Wash the filter using 5% TCA, and then 1% TCA in the same way.
- Add 4.5 ml Pico-Fluor™ 15, vortex and count.

12.3.4 Results

From the counts (n dpm), determine the specific activity of DHAP-AT in the samples (nmoles of acyl-[^{14}C]DHAP formed/mg protein/min).

Activity = n dpm / (specific activity in dpm/nmol) x (1000 / 10 µg protein) x 1.5 * (* for compensating for the 200 µl aliquot taken from the 300 µl chloroformic phase)

DHAP-AT activity was measured at pH 5.4 in homogenates of rat hepatoma cells (MH_1C_1) treated with 250 µM ciprofibrate during several periods of time (see protocol 11). For comparisson, the DHAP-AT activity in liver homogenates and purified peroxisomes is reported.

Samples	Acyl-DHAP formed (nmol/min/mg protein)
Control MH_1C_1 (24 h)	0.348 ± 0.02
Control MH_1C_1 (48 h)	0.386 ± 0.015
Control MH_1C_1 (72 h)	0.311 ± 0.004
250 µM Ciprofibrate treated MH_1C_1 (24 h)	0.472 ± 0.01
250 µM Ciprofibrate treated MH_1C_1 (48 h)	0.443 ± 0.03
250 µM Ciprofibrate treated MH_1C_1 (72 h)	0.500 ± 0.01
Rat liver homogenate	0.182 ± 0.01
Rat purified peroxisomes	2.495 ± 0.127

Data evidence that the DHAP-AT activity in the MH_1C_1 cell line was little increased by ciprofibrate-treatment, whereas the palmitoyl-CoA oxidase activity was increased four times under the same conditions (see protocol 11 and ref [1]).

Measurement in transformed or not transformed fibroblasts from patients suffering from peroxisomal deficiency diseases showed lower DHAP-AT activities (about 0.12 nmol/min/mg protein, see protocol 13). These activities were restored in hybrid cells obtained after fusion with rat hepatoma cells (about 0.57 nmol/min/mg protein, see protocol 14).

References

1. Brocard C, Es-souni M, Ramirez LC, Latruffe N, Bournot P (1993) Stimulation of peroxisomal palmitoyl-CoA oxidase activity by ciprofibrate in hepatic cell lines: comparative studies in Fao, MH_1C_1 and HepG2 cells. Biol Cell 77:37-41
2. Causeret C, Bentejac M, Clemencet MC, Bugaut M (1993) Effects of two peroxisome proliferators (ciprofibrate and fenofibrate) on peroxisomal membrane proteins and

dihydroxyacetone-phosphate acyl-transferase activity in rat liver. Cell Mol Biol 39:65-80

3. Hajra AK, Bishop JE (1982) Glycerolipid biosynthesis in peroxisomes via the acyl-dihydroxyacetone phosphate pathway. Ann NY Acad Sci 386:170-182

4. Hajra AK, Burke CL, Jones CL (1979) Subcellular localization of acyl Coenzume A: dihydroxyacetone phophate acyltransferase in rat liver peroxisomes (microbodies). J Biol Chem 254:10896-900

5. Hardeman D, van den Bosch H (1989) Topography of ether lipid biosynthesis. Biochim Biophys Acta 1006:1-8

6. Hardeman D, Zomer HWM, Schutgens RBH, Tager JM, van den Bosch H (1990) Effect of peroxisome proliferation on ether phospholipid biosynthesizing enzymes in rat liver. Int J Biochem 22:1413-1418

7. Schutgens RBH (1984) Deficiency of acyl-CoA: dihydroxyacetone phosphate acyltransferase in patients with Zellweger (cerebro-hepato-renal) syndrome. Biochem Biophys Res Commun 120:179-184

8. Singh H, Usher S, Poulos A (1989) Dihydroxyacetone phosphate acyltransferase and alkyldihydroxyacetone phosphate synthase activities in rat liver subcellular fractions and human skin fibroblasts. Arch Biochem Biophys 268:676-686

9. Hajra AK, Jones CL, Davis PA (1978). In Freys L, Dreyfus H, Massarelli R, Gatt S (eds) Enzymes of lipid metabolism II, New York, Plenum, pp. 369-378

10. Van den Bosch H, Schutgens RBH, Wanders RJA, Tager JM (1992) Biochemistry of peroxysomes. Annu Rev Biochem 61:157-197

11. Wanders RJA, van Roermund CWT, Griffioen M, Cohen L (1992) Peroxisomal activities in the human hepatoblastoma cell line HepG2 as compared to human liver. Biochim Biophys Acta 1115:54-59

13 In Vitro Transformation of Human Fibroblasts with SV40 T Antigen (Lipofection)

K. H. NG, E. B. BIOUKAR, F. STRAEHLI and J. DESCHATRETTE

13.1 Introduction and Aims

Transformed established cell lines are a useful material for molecular genetic investigations of heritable diseases. Indeed, experiments involving the search for correlations between modifications of the genetic material of the cells (fusion, transfection) and phenotypic transitions require long term culture, selection of clones and subclones, and cell characterization, which are difficult to perform with untransformed cells. Furthermore, besides the slow growth rate of untransformed cells, the senescence occuring after a number of cell generations limits the potential investigations performed on usually scarce and unique material. This is especially true for samples obtained from patients suffering from peroxisomal diseases. The still unexplained wide range of clinical and biochemical cases render each sample potentially precious.

Considering the relatively low traumatic sampling, two kinds of cells may be obtained in a current way from patients: lymphocytes (blood samples) and fibroblasts (skin biopsies). Both cell types may be transformed *in vitro* and exhibit respective advantages and disadvantages for experimental work. The ability of the fibroblasts to adhere to plastic support and grow in monolayers eases the clonal analysis of transformed material.

The transformant potential of SV40 T antigene on human fibroblast led to the protocol described below [3, 4, 5, 6, 7]. It exploits the methods used to introduce foreign genetic material in mammalian cells (transfection). Several transfection techniques have been developped, including the uptake of DNA as calcium phosphate-DNA precipitate, electroporation and lipofection. The choice for lipofection is based on its technical simplicity and efficiency in our experimental context. Lipofectin reagent is a 1:1 (w/w) liposome formulation of the cationic lipid N-{1-(2,3-dioleyloxy)propyl}-N,N,N-trimethylammonium chloride (DOTMA) and dioleoyl phosphatidylethanolamine (DOPE) in water. The lipofectin complex and entrapped DNA fuse with the cell membrane, resulting in DNA uptake [2].

In the following experiment, the transformant used is pAS, which is a pBR322 plasmid in which was inserted a deletion mutant of the SV40 early gene fragment [1].

13.2 Equipment, Chemicals, Culture Medium and Solutions

13.2.1 Equipment

13.2.1.1. Access to

- a 10% CO_2/90% air humidified incubator (Jouan series EB)
- a vertical laminar flow hood with a vacuum pump and gas
- a low speed centrifuge (Jouan B 3.11)
- a phase contrast microscope (Olympus CK 2 TR)
- a binocular low magnifier (Olympus)
- a 37°C water bath
- adjustable automatic pippettes (Gilson or Labsystems) with sterile tips
- automatic portable pipetting aid devices (Sarstedt)

13.2.1.2 Disposable Plastic Material

- sterile plastic pipettes(1, 2, 5 and 10 ml)
- plastic dishes for cell culture (3.5, 6.0 and 10.0 cm in diameter)
- 15 ml polystyrene conical tubes (Falcon 2095)
- unplugged and plugged Pasteur pipettes (CML)
- cryotubes for cell conservation (CML)
- 0.2 µm membrane filters (Schleicher and Schuell FP 030/3)
- forceps (Millipore)
- a Neubauer cell counting slide

13.2.2 Chemicals

- penicilline (Eurobio)
- fungizone (Squib)
- streptomycine (Eurobio)
- Lipofectine reagent (1mg/ml, Gibco BRL 8292 SA)
- dimethylsulfoxide (DMSO; Sigma)
- pig pancreas trypsin (U.S.B. 22715)
- NaCl, KCl, anhydrous dextrose, EDTA (Na_2), Na_2HPO_4 ($2H_2O$) and KH_2PO_4 (molecular biology quality, Merck)
- sterile bidistilled water

13.2.3 Culture Medium and Solutions

- standard culture medium: Ham F12 medium modified by Coon (Gibco-BRL SP244A) containing 5 % fetal calf serum (Institut Jacques Boy) and antibiotics PFS (1 X) (see below)
- antibiotics PFS (100X): dissolve 200,000 U penicilline, 5mg fungizone and 0.5 g streptomycine in 100 ml sterile PBS solution (see below). Store as 10 ml aliquots at -20°C.
- 0.05% trypsin solution (ATV): dissolve 8.0g NaCl, 4.0g KCl, 10.0g anhydrous dextrose, 2.0g EDTA, Na$_2$, and 0.50g pig pancreas trypsin in 1 L of bidistilled water, and incubate at 37°C for 1 h. Filter with a 0.2 μm membrane. Freeze at -20°C as 50 ml aliquots.
- saline phosphate buffer (PBS): dissolve 8.0g NaCl, 2.0g KCl, 1.44g Na$_2$HPO$_4$ (2H$_2$O) and 2.0g KH$_2$PO$_4$ in 1 L of bidistilled water, and filter with a 0.2 μm membrane.

13.3 Experimental Procedure

13.3.1 Preparation of Cell Monolayers

Fibroblasts grown after the establishement of primary cultures from skin biopsies are the biological material of the transfection experiments. Vials of cells frozen in liquid nitrogen or in a -80°C freezer are rapidly thawed in a 37°C water bath with constant and gentle agitation. The cell suspension is added to 2 ml of culture medium warmed at 37°C and spun at 500g for 3 min. The supernatant is discarded. The cell pellet is gently suspended in culture medium and placed in the incubator. The next day, the medium is renewed and the cells are further grown to confluency.

Day 1

The healthy state and confluency of the cell culture is verified with a phase contrast microscope. The medium is removed and 5 ml of ATV (trypsin containing solution) are added to the cell monolayer in the culture dish. The dish is then incubated at 37°C for 2 to 3 min. The cells, now submerged in ATV, are detached from the culture dish by gentle pipetting and transferred to a 15 ml conical plastic tube containing 2 ml of culture medium. After centrifugation (500g, 3 min., room temperature), the supernatant is discarded, and the cells are suspended in culture medium, counted and reinoculated in 10 cm dishes at a density of 1.5 x 10^6 cells/dish with 10 ml of medium.

Day 2

The culture medium is removed, the cell monolayer is washed 3 times with serum-free medium and 5 ml of serum-free medium are added per 10 cm dish..

13.3.2 Lipofection

After ethanol precipitation of 10 µg of plasmid DNA, the pellet is resuspended in 50 µl of sterile water and diluted with serum-free medium to a volume of 0.5 ml. Thirty µg of Lipofectin reagent are also diluted to 0.5 ml with serum-free medium. The DNA and lipofectin solutions are then mixed in a polystyrene tube and allowed to stand for 15 min at room temperature. Finally, the mixture is added to the cell monolayer and then incubated at 37°C for 5 to 24 hours.

The culture medium is then replaced with that containing 5% of serum.

Day 3

The cells are harvested with ATV solution, as previously described, and reinoculated into 4 dishes with standard medium. The cultures are grown further for 4 to 6 weeks with medium renewal every 3 days. Cellular foci of multilayers can then be observed, picked up and expanded.

13.3.3 Obtention of Immortalized Strains

Foci of transformed cells have been isolated from cells transfected one month before as described above. The transformed clones may be identified as white fuzzy spots in the dishes. Under a phase contrast microscope, round cells undergoing mitosis may be observed in these cellular foci in multilayers. These clones are picked up by aspiration using capillaries made from cotton-plugged Pasteur pipets, inoculated into 3 cm culture dishes and expanded. Ultimately, the clones may be characterized for the presence of plasmid DNA or for phenotypic properties. Aliquots of these clonal cells are then frozen in liquid nitrogen.

As described above, the transformed clones are identified on the criterion of loss of contact inhibition of growth and a lower requirement for serum (see comments). These transformants are further characterized by a faster growth rate (approximately 24 hours per cell generation) and a longer life span. Some clones may develop as permanent, immortalized strains. Yet, this is not the rule. Usually, after 2 to 4 months of continuous growth, the cultures enter a "crisis" characterized by growth arrest and cell death. Sub-clones may appear in these cultures that will develop into permanent strains.

13.3.4 Cell Freezing

Subconfluent cultures are harvested by trypsin treatment, as described above. Cells are counted and centrifuged, and the pellet is resuspended in medium containing 5% fetal calf serum supplemented with 10% DMSO at a concentration of 5×10^5 to 10^6 cells/0.5 ml. Aliquots (0.5 ml) are distributed in sterile plastic cryotubes and frozen in either liquid nitrogen or at -80°C. Whereas the thawing of cell inoculations requires a sharp rise in temperature (see above), the freezing to -80°C or to -195 °C (liquid nitrogen) must be progressive. This can be performed by wrapping the tubes in thick paper followed by freezing at -80 °C for 24 hours. The tubes can then be transferred either into storage boxes at -80 °C or into liquid nitrogen. Some liquid nitrogen containers are equiped with systems for gradual freezing of the material.

13.4 Comments

The tranformation efficiency (frequency of transformants) is highly dependant on the growth potential of the fibroblasts at the time of transfection. It depends on the source of the biopsy (fetus or adult) and on the rate of cell generation in vitro. The transformation of strains close to senescence is discouraged.

The experiment requires long term culture of the cells to confluency. Although non-transformed cells stop dividing under these conditions, further incubation may lead to cells detaching from the plastic support. This can be counteracted, first by cell inoculation at low densities, and second by reducing the amount of serum in the medium. After confluency is reached (8 to 10 days), cells are cultured in medium containing 0.5% to 0.1% of serum for 10 more days. As the serum requirement for transformants is lower than that for non-transformed cells, the emergency of clones is not affected.

References

1. Benoist C, Chambon P (1981) In vivo sequence requirements of the SV40 early promoter region. Nature 290:304-310
2. Felgner PL, Gadek TR, Holf M, Roman R, Chan HW, Wenz M, Northrop JP, Ringold GM, Danielson M (1987) Lipofection: a novel, highly efficient lipid mediated DNA transfection procedure. Proc Natl Acad Sci USA 84:7413-7417
3. Girardi AJ, Jensen FC, Koprowski H (1965) SV40-induced transformation of human diploid cells: crisis and recovery. J Cell and Comp Physiol 65:69-84
4. Huschtscha LI, Holliday R (1983) Limited and unlimited growth of SV40-transformed cells from human diploid MRC-5 fibroblasts. J of Cell Sci 63:77-99

5. Mayne LV, Priestley A, James MR, Burk JF (1986) Efficient immortalization and morphological transformation of human fibroblasts by transfection with SV40 DNA linked to a dominant marker. Exp Cell Res 162:530-538
6. Neufeld DS, Ripley S, Henderson A, Ozier HL (1987) Immortalization of human fibroblasts transformed by origin-defective simian virus 40. Mol Cell Biol 7:2794-2802
7. Wright WE, Pereira-Smith OM, Shay JW (1989) Reversible cellular senescence: implications for immortalization of normal human diploid fibroblasts. Mol Cell Biol 9:3088-3092

14 Somatic Cell Hybridization as a Tool for Genetic Analysis of Peroxisomal Activities

E. B. BIOUKAR, F. STRAEHLI, K. H. NG and J. DESCHATRETTE

14.1 Introduction and Aims

Since the initial observations in 1960 and the methodological developments that followed, somatic cell hybridization has been a fruitful tool for genetic analysis in numerous domains (see [4] for theory and [7, 8] for methodology). This technique may be exploited in the investigations of inherited peroxisomal diseases and of gene regulation related to the organelle. The determination of recessive and dominant traits of mutations, complementation analysis and the regulation of peroxisomal activities in relation to cell differentiation may be undertaken with this method. Basically, such analysis can be performed on early products of fusion (mostly heterokaryons), or alternatively, if one of the parents is a permanent cell line, on growing hybrid clones.

The studies done on heterokaryons have the advantage of short term analysis and are adequate for complementation analysis, as well. Such analyses of early products of fusion can be carried out from a few hours to a few days after fusion. It may be accompanied by in situ analysis (*e.g.* immunocytochemisty), by the detection of enzymatic activies in the analysis of cell populations (heterokaryons + unfused cells), or by the search for specific transcripts. Growing hybrid clones permit the establishment of correlations between the genetic complement of the cells and phenotypic traits. Such an approach exploits chromosome segregation, which occurs in the off-springs of the fusion products, leading to genetic heterogeneity of the hybrid clone. Thus, the study of growing hybrids requires selective systems of fusion products.

Two series of fusions are detailed below. The aim of the first experiment is a complementation analysis between two cell lines established from patients with Zellweger syndrome and neonatal adrenoleukodystrophy (NALD), respectively, both deficient in peroxisomal activities [1, 2, 3, 5, 6, 9, 10, 11]. The experiment consists of the construction and early phenotypic characterization of heterokaryons between untransformed fibroblasts of the two lines.

The second fusion experiment is a genetic approach to the regulations that dictate the cell-specific level of expression of peroxisomal functions. Indeed, although ubiquitous, peroxisomes vary significantly in number, size and structure, depending on the cell type. While liver cells exhibit numerous organelles and high enzymatic activities, fibroblasts are characterized by much fewer peroxisomes and lower enzyme activities. Whether or not such differences reflect regulations coordinated with the control of expression of differentiated functions is unknown. The extinction of cell-specific traits is a general rule in somatic hybrids. The investigation of the regulation of peroxisomal functions starts with the characterization of the hybrids between two cell types possessing different levels of activities. The fusion between rat hepatoma cells and human fibroblasts will be performed, leading to the selection of growing hybrid clones.

14.2 Equipment, Chemicals, Culture Medium and Solutions

14.2.1 Equipment

As in Exp. 13

14.2.2 Chemicals, Culture Medium and Solutions

As in Exp 13. Specific chemicals, medium and solutions for cell fusion and selection are detailed:
- HAT medium (100X): 10 mM hypoxanthine, 40 μM aminopterin and 1.6 mM thymidine. Dissolve each component (136 mg, 1.76 mg and 38.8 mg respectively, Sigma) in bidistilled water with a few drops of 1N NaOH and then mix them together. Add water to make a final solution volume of 100 ml and filter with a 0.2 μm membrane. Aliquot the HAT (100X) solution and freeze at -20° C.
- 0.1M ouabain (100X): heat 2 g of ouabain (Sigma) in 27.44 ml of bidistilled water to dissolve it and then filter with a 0.2 μ membrane. Freeze the aliquots at -20° C.
- polyethylene glycol (PEG) 1000 (Merck)

14.2.3 Biological Material

- untransformed fibroblast cell lines from Zellweger and NALD biopsies
- transformed human fibroblasts
- rat hepatoma cells (H4IIEC3; Reuber H35 hepatoma). These cells are characterized by two genetic markers: HGPRT⁻ (conferring the resistance to 8-azaguanine and sensibility to HAT medium) and ouabain^R (resistance to 2 μM ouabain)

14.3 Experimental Procedure

Day 1

14.3.1 Inoculation of Mixed Parental Cells

14.3.1.1 Fusion 1: Zellweger Syndrome x NALD (Untransformed Fibroblasts)

The state of the cell cultures is checked under a phase contrast microscope. The medium is removed and 5 ml of ATV (trypsin containing solution) are added to the cell monolayer. The flask is then incubated at 37 °C for 2 to 3 mins. Cells, now submerged in ATV, are detached from the culture dish by gentle pipetting and transferred into a 15 ml conical plastic tube. Two ml of culture medium containing 5% fetal calf serum are added to the cell suspension. After centrifugation (500g, 3 min, room temperature), the supernatant is discarded, and the cells are resuspended in culture medium and counted. A mixture of 10^6 cells of each line is added in a series of 10 cm dishes, 7 ml of medium are added per dish, and the dishes are then transferred to the incubator.

14.3.1.2 Fusion 2: Transformed Human Fibroblasts x Rat Hepatoma Cells

The two types of cells are harvested under the same conditions, as described for the first fusion. The two different cell types are mixed at a concentration of 1:1. Three series of dishes are inoculated with the following concentrations of cells, respectively:

$1 \cdot 10^6$ of human fibroblasts x $1 \cdot 10^6$ rat hepatoma cells
$2 \cdot 10^6$ of human fibroblasts x $2 \cdot 10^6$ rat hepatoma cells
$3 \cdot 10^6$ of human fibroblasts x $3 \cdot 10^6$ rat hepatoma cells.

The dishes are then transferred to the incubator.

14.3.2 Treatment with a 50% PEG Solution

PEG 1000 is liquified in a 65° C water bath. The desired weight of liquified PEG 1000 is poured into a Pyrex glass bottle and sterilized for 30 min in the autoclave. The sterile serum-free medium is warmed to 65° C and the required volume of medium is added to the sterile PEG 1000 (10 ml/10g). Five ml of the solution will be needed for the treatment of each dish. The 50% PEG 1000 solution can then be stored at 4° C.

Day 2 *(Morning)*

The series of dishes corresponding to the two fusion experiments are treated in the same way.

The medium is removed and the dishes are washed 3 times with 10 ml of serum-free medium. Five ml of the 50% PEG 1000 solution, prewarmed to room temperature, are added on the cell monolayer. After 45 seconds of treatment, the mixture is removed, and the cells are washed with 5 ml of serum-free medium. The washes are repeated several times.

The cells are rendered very fragile after the PEG treatment. Thus, great care must be undertaken during the washes. The medium from the pipette must be directed against the edge of the dish to avoid cells detaching from their plastic support. Finally, 10 ml of medium containing serum are added, and the dishes are transferred to the incubator.

Day 2 *(Late Afternoon)*

Fusion 2: The cells are detached using ATV, counted and re-inoculated at various densities: 10^5, 5.10^5 and 10^6 cells/10 cm dish. Medium containing serum is then added to the dishes, which are transferred to the incubator.

Day 3

Fusion 1: The PEG treated cells are allowed to recuperate several hours before the cultures are harvested for analysis by in situ immunocytofluorescence, enzymatic dosage or RNA transcription activities.

Fusion 2: The medium is removed and replaced with selective medium containing HAT and ouabain. Two weeks later, the clones of selected hybrids will be identified and picked up from the dishes. These clones are picked up by aspiration using capillaries made from cotton-plugged Pasteur pipettes inoculated into 3 cm culture dishes and expanded. Ultimately, the hybrids can be characterized for their phenotypic properties in parallel with karyological analyses. Aliquots of the clones are frozen in liquid nitrogen.

References

1. Allen LA, Morand OH, Raetz CRH (1989) Cytoplasmic requirement for peroxisome biogenesis in Chinese hamster ovary cell. Proc Natl Acad Sci USA 86:7012-7016
2. Brul S, Wiener EAC, Westerveld A, Strijland A, Wanders RJA, Schram W, Heymans HSA, Schutgens RBH, van den Bosch H, Tager JM (1988) Kinetics of the assembly of peroxisomes after fusion of complementary cell lines from

patients with the cerebro-hepato-renal (Zellweger) syndrome and related disorders. Biochem Biophys Res Comm 152:1083-1089

3. Brul S, Westerveld A, Strijland A, Wanders RJA, Schram AW, Heymans HSA., Schutgens RBH., van den Bosch H, Tager JM (1988) Genetic heterogeneity in the cerebrohepatorenal (Zellweger) syndrome and other inherited disorders with a generalized impairment of peroxisomal functions: a study using complementation analysis. J Clin Invest 81:1710-1715

4. Ephrussi B (1972) Hybridization of somatic cells. Princeton Univ Press Princeton, New Jersey

5. McGuinness MC, Moser AB, Moser HW, Watkins PA (1990) Peroxisomal disorders: complementation analysis using beta-oxidation of very long chain fatty acids. Biochem Biophys Res Comm 172:364-369

6. Poll-The BT, Skjeldal OH, Stokke O, Poulos A, Demaugre F, S Saudubray JM (1989) Phytanic acid alpha-oxidation and complementation analysis of classical Refsum and peroxisomal disorders. Hum Genet 81: 175-181.

7. Pontecorvo G (1975) Production of mammalian somatic cell hybrids by means of polyethylene glycol treatment. Somatic Cell Genet 1:397-400

8. Ringertz NR, Savage RE (1976) Cell hybrids. Acad. Press, New York

9. Roscher AA, Hoefler S, Hoefler G, Paschke E, Paltauf F, Moser A, Moser H (1989) Genetic and phenotypic heterogeneity in disorders of peroxisome biogenesis-a complementation study involving cell lines from 19 patients. Ped Res 26:67-72

10. Singh AK, Kulvatunyou N, Singh I, Stanley WS (1989) In situ genetic complementation analysis of cells with generalized peroxisomal dysfunction. Hum Hered 39:298-301

11. Yajima S, Yasuyuki S, Shimozawa N, Yamaguchi S, Orii T, Fujiki Y, Osumi T, Hashimoto T, Moser HW (1992) Complementation study of peroxisome-deficient disorders by immunofluorescence staining and characterzation of fused cells. Hum Genet 88:491-499

N. LATRUFFE and S. J. HASSELL

PEROXISOMES, Biochemistry, Molecular Biology and Genetic Diseases - A Video Film

So far, peroxisome proliferation in rodent liver has proven to be a good marker of toxicology for several classes of xenobiotics, including fibrates, used as hypolipemic drugs; phtalates and adipates, as plasticizers; or chlorophenoxy-acetate, as herbicide.

Research in peroxisomes provides a good example of the integration of fields related to basic sciences and biomedical and industrial health. Indeed peroxisomes are now condidered as toxicological indicators for several classes of xenobiotics. The subject of peroxisomes has explored during the past few years, becoming ever more attractive in cell biology and pathology.

A 25 min video cassette displays techniques involving the characterization of purified peroxisomes and membranes, immunoblotting, measurement of proliferation markers, mRNA analysis at the post-transcriptional level, DNA techniques, cell cultures as biological models and computer analysis.

This VHS film has been registered at the french Bibliotheque Nationale under the reference n° DV 93 02736.